Excel 2021

A Fool-Proof Guide to Master the Fundamentals of Excel Grasping Advanced Features like Business Modelling, Sampling Design and Data Analysis Techniques

WEBSTER TYLER

clarifying purposes only and are owned by the owners themselves, not affiliated with this document.

Table of Contents

Introduction

When we talk about Excel, there isn't much room for compromise. You have a lot of happy clients, and you're going to be singing the tracks on a spreadsheet for most of the day. On the other side, there are many others who oppose it. They believe it is a row-and-column software program, but advanced Excel theories concentrate on a diverse range of fundamental skills that can be applied and recognized in virtually any position within an organization. Because you've learned these concepts, you should be more knowledgeable about:

- Make sense of the data by visualizing, analyzing, and evaluating it.
- Create spreadsheets to aid in the coordination of details and provide a clearer view of the incoming data.
- Create equations that give you more information about crucial business functions like workflow, project efficiency, financial estimates, and budget projections. It aids in inventory levels and consumption management.
- Maintain a healthy balance of financial and entire inventory levels records
- Create an easy-to-use data collection tool for senior managers to use when assessing ongoing activities or conditions in an entire organization.
- Create reporting systems for a variety of agencies and operations, using various workflow procedures.
- Analyze and comprehend data from other authorities, vendors, and customers.
- Allows consumers to interpret data in a more sophisticated way, providing responses and feedback to business concerns.
- Advanced Excel training will provide businesses with highly skilled employees. Workers would be able to perform more effectively in their current positions and move to relatively higher positions as a result of it.

In this millennium, the most widely used spreadsheet program is Microsoft Excel. Many important features are included in the latest MS Excel (2016 to 2021), allowing you to fully discover the power of Excel.

- For planners and accountants, this program is unrivaled.
- Make the work go more quickly.
- Teach you all aspects of data analytics, including data storage and model execution.
- The most widely used and applicable spreadsheet for business purposes by increasing efficiency and production.
- To establish yourself as an essential part of the company.

Just to simplify, it's always believed that learning is a continuous process, and there's no better way to empower yourself than through honing your skills and the value of your company through information and innovative technologies.

Chapter 1: Seeking to learn Microsoft Excel

Microsoft Excel is a software spreadsheet developed by Microsoft for the Windows operating system, iOS operating system, Mac operating system, and Android operating system. Estimation, interactive processes, including pivot tables, are all used. Visual Basic, however, is a macro programming language that comes with it. This whole section of the book aims to provide detailed information on Microsoft Excel 2021.

1.1 What exactly is Excel?

A table can be represented by a sequence of rows & columns in the Excel spreadsheet software. Generally, alphabetical characters are assigned to columns, while numerical characters are assigned to rows. A cell is a point whereby a column and one row meet. The letters at the top of a column and the numbers that mark the rows tend to indicate the address of almost any cell.

1.2 Versions of Excel

Since 1985, nearly 29 versions of Excel are being published, ranging from 1985 here to the current date. The majority of people today have either 2016, 2019, or Office 365 (2021), the latest version of Excel. Almost every Excel edition was vastly unique from the previous one, just like today's Mac and MS Windows operating systems. A list of Microsoft Excel releases can be found below.

Versions of OS/2 (2.2, 2.3, and 3): In 1985, Microsoft and IBM collaborated on the OS/2 operating system. In 1992, it was solely acquired by IBM, but three models of Excel were launched for OS/2 in the interim.

Mac older versions (1, 2, 3, 4, 5, 98,2k and 2k1)

While Microsoft had a much early spreadsheet product named Multiplan that was featured on MS-DOS and other console-based operating platforms, it's less known how the first edition of Excel was only introduced on the Mac.

Excel's first Windows version was originally a port of Mac's "Excel 2."

Although Microsoft recently launched for Windows and MAC the Excel 2019 under such a similar name, they did so previously with Excel 2k, which was available for Mac and Windows.

Windows Older versions (2.0, 3.0, 4.0, 95, 97, 2k and 2k2)

Excel 2.0, which had been released in 1987, is one of the earliest versions of the program. Some of Excel's most well-known features date back a long time!

Excel 4.0, which had been released in 1992, was the first version to have AutoFill.

VBA and Macros have been around since Excel 5.0 was published in 1993. Excel became a significant target for macro viruses due to the versatility of VBA before Excel 2007 updated the formats of a file to improve protection.

In Excel 97, "Clippy," the Office Assistant was added; however, most people found it highly irritating! In Excel 2002, this was turned off by default, and in Excel 2007, it was completely disabled.

Windows (Microsoft Excel 2k3)

Excel 2003 is a spreadsheet program (Windows)

Excel 2003 has been the last version of Excel to be using the old "WIMP" (Windows, Icons, Menus, and Pointer) GUI. You could recall the drop-down icons & menus at the top edge of the screen if you're using them.

Excel 2003 has been the first edition of Excel to have the Tables feature, which was later greatly enhanced.

Mac (Excel 2K4)

Excel 2004 had been available for only Apple Macintosh computers.

Windows (Excel 2k7)

Excel 2007 was also only available for PCs running Windows.

Excel 2007 had been a significant improvement over previous versions, adding the Ribbon interface and switching the format of File from the common.xls to the latest .xlsm and .xlsx formats. Excel files could now hold around 1 million rows (instead of the previous cap of 16,384), and security was greatly improved. The charting features of Excel have also been greatly enhanced in this version.

Many users disliked the new interface all at the time, according to surveys, but Microsoft kept with it, and most users today might not like to go back.

Mac (Excel 2k8)

Excel 2008 was also only available on Mac Apple computers.

Windows (Excel 2k10)

Excel 2010 was also only available for PCs running Windows. Multi-threading support, Ribbon customization, back-stage view, and spark lines are among the latest features.

Despite the fact that Excel 2010 was the 13th version of the software, Microsoft omitted version 13 then announced it version 14 because the number 13 is considered unlucky.

Mac (Excel 2k11)

This was the last Mac version of Excel to have a different name from the Windows version.

Excel 2011 was only available on Mac computers. This was the last Mac version of Excel to have a different name from the Windows version.

This too has been the 13th edition of Mac Excel, much like the Excel 2010 and Windows; however, version 13 was dropped for superstitious grounds, whereas Excel 2011 was announced to have been version 14.

Windows (Excel 2k13)

Excel 2013 launched the latest Slicers, 50 new functions, and Flash Fill feature, but it was only available for Windows computers.

Excel 2k16 & MS 365 for windows

For maybe the first time until 2000, Microsoft agreed to offer the Mac & Windows versions of Excel the same name. Even after this, these two programs are somewhat different, owing in part to the iOS operating system's different user interface.

Rather than updating new features whenever a whole new version is launched, Microsoft started to release daily feature updates through the internet with Excel 2016. Users of packaged retail versions of Excel now have a somewhat different version of Excel than Office 365 users since these latest features were available only to subscribers of Office 365.

Excel 2k19

The incorporation of "power" tools: Power Pivot, Power 3-D Maps, and Power Query (Get & Convert), among all the versions, was a significant improvement for Excel 2019. These were all high-level tools (OLAP) that enabled Excel to evaluate "big data" and perform "latest data analysis" through any Excel version. However, since the introduction of

Excel 2019, the "power" tools have expanded further, with additional "power" features applicable only to users of Excel 365.

Excel 365

Excel 365 is designed to have an eternal existence, much as there may not be Windows 11. Instead of the traditional three-year upgrade cycle, 365 is an ever-evolving product, including new features added with each new edition. Microsoft pioneered the idea of "update channels" now for Office 365 as business customers don't want continuous changes and want rock-solid, stable releases). Users in the business world should sign up for a six-month update channel. In January & July of each year, this produced a strong and checked "Semi-annual" version.

Microsoft hasn't revealed all of Office 2021's features and updates yet; however, the Office (Long-Term Servicing Channel) LTSC version would have accessibility improvements, dark mode support, and features including XLOOKUP and Dynamic Arrays. Similar features will be available in Office 2021.

Visually, dark mode is by far the most noticeable update, but Microsoft will continue to prioritize the Microsoft 365 variants of Office for the majority of its interface and cloud-based features.

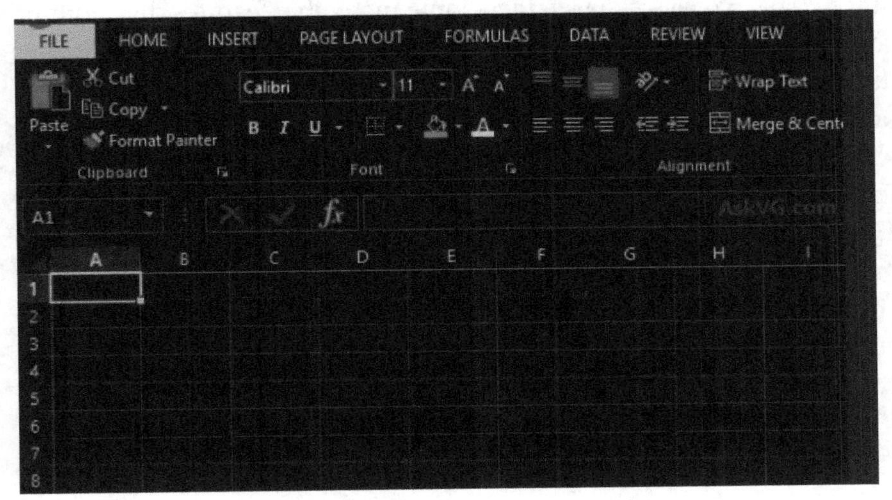

Many earlier Excel functions can be used in entirely different ways in Excel 365, and also some old classics have also been updated with new array-aware functions. VLOOKUP, for example, has been substituted by XLOOKUP.

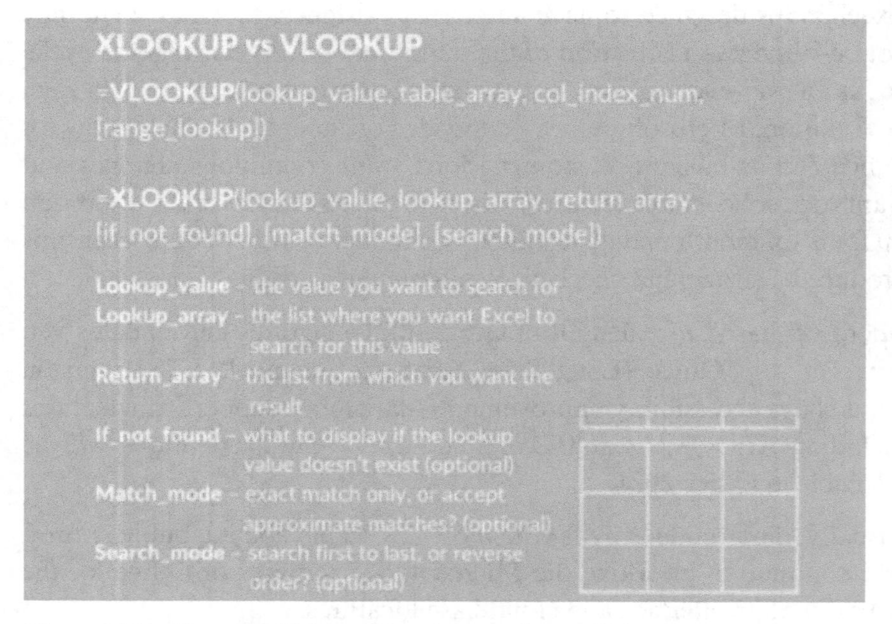

Since 2019, Excel 365 has gained some incredible features, but nothing compares to the launch of Dynamic Arrays & Dynamic Array functions during the July 2020 version release. This is not a minor adjustment. In order to accept the concept of the latest dynamic array, the Excel engines were to be re-engineered. Excel 365 has advanced to a different level than Excel 2019, thanks to the latest dynamic array capabilities. Dynamic arrays are incompatible with Excel 2019, which therefore will enable modern workbooks (generated with Excel 365) to fail to work as anticipated in older versions (Excel 2019 in this context is a legacy version, although one may still purchase it).

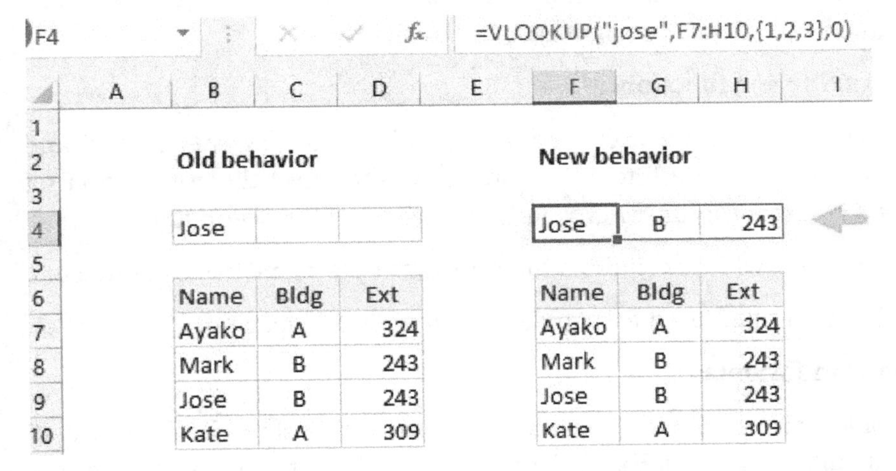

Excel 365 is by far the most recent, best, and perhaps most potent excel edition available, and it's available for a low monthly fee.

1.3 Most Common Features

Excel provides its consumers with a wealth of features that they can tailor to their specific requirements. Listed here are a few of the most common functions that otherwise excel claims to have.

Adding a Header and a Footer

In the File of the spreadsheet, Microsoft Excel assists us in maintaining the header and footer parts.

Password-protected files

It allows users to protect their workbooks from unauthorized access and share information through a code or password.

Replace & Find

MS Excel assists one in finding the required data (text or numbers) in their workbook. One can also replace the original old data with new information.

Commands of Data Sorting

Data sorting in Excel is defined as a manner of grouping the data in such a sequential order. Users can sort data both in descending &ascending sequence, and vice versa, using MS Excel.

Data filtering function

Filtering data inside a collection is a quick and easy way to find and handle it. Within a filtered list, just rows that meet the criteria you set for a column appear. Microsoft Excel has two filtering options:

AutoFilter, which includes a filter through option and basic parameters.

Advanced filter; used for standards that are more specific.

Built-in formulas

In addition, minimum, average, maximum, and other formulas are all built into Microsoft Excel. You may use these formulas to meet the requirements.

Counting a specific set of data

In Excel, there are three main functions that can be used to count every row or column.

- **Count Function (start Address: end Address):** This feature is being used to count the number of numerical values such as 1,2,3,4, 5, up to infinity.
- **Count A Function (start Address: end Address):** This feature counts alphanumeric characters for the consumer. It may be anything from A to Z, a to z, or 0 to 9 unique characters.
- **Count Blank Function (start Address: end Address):** This feature is used to determine how many blank cells are available in a given row or column.

Validation of Data

Data validation plays a role whenever we would not like a user to input any adulterated data into the excel sheet. There are numerous kinds of data validations available, including the following:

- The database for a whole number is primarily validated by whole number validation.

- Floating values are the main subject of decimal validation.

1.4 Five Productivity-Boosting Excel Features

Excel's Ideas feature will help you automate data analysis

An idea is sort of an artificial intelligence feature in Excel that is compatible with Office 365 subscriptions. Excel will quickly analyze the data with Ideas and provide you with information you would not have found otherwise. Ideas may be helpful in different cases, for example:

- Rank data and classify objects that are substantially smaller or larger than for the majority of the people by analyzing transactions;
- Trend analysis is used to illustrate patterns in data that have developed over time.
- Detecting significant outliers throughout data points, such as transactions that may be erroneous or fraudulent; .besides
- Calling attention to circumstances where a large portion of the overall value is due to a single cause.

You can view Ideas from the Home button of the Ribbon if you have the Office 365 subscription. However, for using this feature, you should have an adequate internet connection.

MAXIFS, IFS, and MINIFS are three conditional formulas that are easier to use

You can construct formulas that involve several tests quite easily than before with MAXIFS, IFS, and MINIFS. Most Excel users used to "nest" numerous IF functions in about the exact formula before IFS became available. When you needed to make a calculation based on one or more conditions, this was standard practice. However, since the advent of IFS, specific formulas have been greatly simplified. For example, note how just one IFS function would be needed in the formula underneath to run three tests on the data within cell A2. This method avoids the need for multiple IF functions, which would have been essential in the past.

=IFS(A2>400,"Tier 1",A2>300,"Tier 2",TRUE,"Tier 3")

MINIFS and MAXIFS, like IFS, allow you to run multiple tests on the data. When using MAXIFS, when all of the tests are passed, Excel would return the highest value. While using MINIFS, on the other hand, when all of the tests are passed, Excel returns the smallest value. Excel 2019 users have access to these features. They're also open to Office 365 subscribers who have access to Excel.

XLOOKUP: A Simpler and Better VLOOKUP Alternative

VLOOKUP and related functions like HLOOKUP and INDEX pale in comparison to XLOOKUP. Although Excel will keep these legacy features, many people will find XLOOKUP to be more transparent and intuitive. Most people will discover that XLOOKUP is much more powerful. The following are some of the main differences among XLOOKUP as well as other lookup functions:

- HLOOKUP & VLOOKUP default to an estimated match, while XLOOKUP defaults to that of an exact fit.
- You don't need to designate any column index number with XLOOKUP like you are doing through VLOOKUP or perhaps any row index number like you would with HLOOKUP.
- With XLOOKUP, the order of rows and columns is irrelevant. That's because, when used as a replacement for VLOOKUP, the feature will look towards the left or right. When used as a replacement for HLOOKUP, it can also look below or above.
- Without adding an IFERROR feature, XLOOKUP makes it possible to determine what happens if the lookup value is not found.

Dynamic Arrays

Another latest feature that is only available with the Office 365 subscription is dynamic arrays. You can use dynamic arrays to compose a single formula that affects several cells at the same time despite needing to copy-paste the formula to each of them. You also don't need to use the CTRL + SHIFT + ENTER keypad pattern to enter a standard array formula if you're using an Excel version that embraces dynamic arrays. Furthermore, if you're using an Excel version that allows dynamic arrays, you'll have access to six additional functions to allow you to take advantage of such a newfound power. FILTER, SORT BY, SORT, UNIQU, RANDARRY, and SEQUENCE are the six functions.

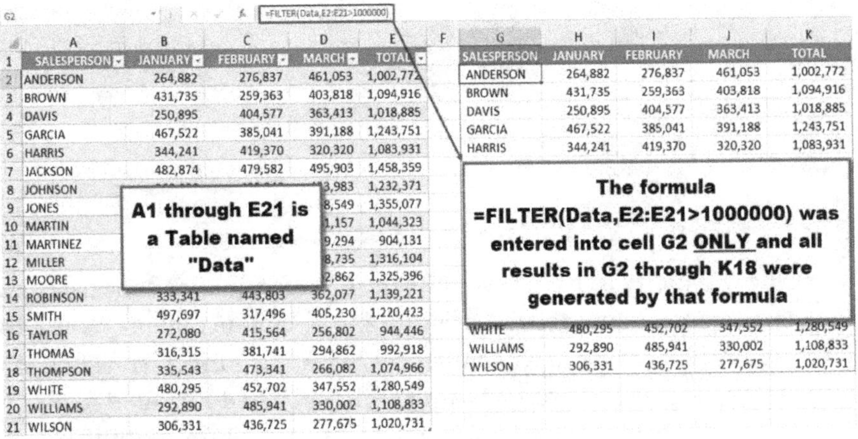

G2 =FILTER(Data,E2:E21>1000000)

	A	B	C	D	E
1	SALESPERSON	JANUARY	FEBRUARY	MARCH	TOTAL
2	ANDERSON	264,882	276,837	461,053	1,002,772
3	BROWN	431,735	259,363	403,818	1,094,916
4	DAVIS	250,895	404,577	363,413	1,018,885
5	GARCIA	467,522	385,041	391,188	1,243,751
6	HARRIS	344,241	419,370	320,320	1,083,931
7	JACKSON	482,874	479,582	495,903	1,458,359
8	JOHNSON	[illegible]	[illegible]	[illegible]3,983	1,232,371
9	JONES	[illegible]	[illegible]	[illegible]8,549	1,355,077
10	MARTIN	[illegible]	[illegible]	[illegible]1,157	1,044,323
11	MARTINEZ	[illegible]	[illegible]	[illegible]9,294	904,131
12	MILLER	[illegible]	[illegible]	[illegible]8,735	1,316,104
13	MOORE	[illegible]	[illegible]	[illegible]2,862	1,325,396
14	ROBINSON	333,341	443,803	362,077	1,139,221
15	SMITH	497,697	317,496	405,230	1,220,423
16	TAYLOR	272,080	415,564	256,802	944,446
17	THOMAS	316,315	381,741	294,862	992,918
18	THOMPSON	335,543	473,341	266,082	1,074,966
19	WHITE	480,295	452,702	347,552	1,280,549
20	WILLIAMS	292,890	485,941	330,002	1,108,833
21	WILSON	306,331	436,725	277,675	1,020,731

A1 through E21 is a Table named "Data"

G	H	I	J	K
SALESPERSON	JANUARY	FEBRUARY	MARCH	TOTAL
ANDERSON	264,882	276,837	461,053	1,002,772
BROWN	431,735	259,363	403,818	1,094,916
DAVIS	250,895	404,577	363,413	1,018,885
GARCIA	467,522	385,041	391,188	1,243,751
HARRIS	344,241	419,370	320,320	1,083,931
WHITE	480,295	452,702	347,552	1,280,549
WILLIAMS	292,890	485,941	330,002	1,108,833
WILSON	306,331	436,725	277,675	1,020,731

The formula =FILTER(Data,E2:E21>1000000) was entered into cell G2 ONLY and all results in G2 through K18 were generated by that formula

With Power Query, you can assess the data quality

The emergence of Power Query, which debuted with Excel 2010's publication, has been nothing less than impressive. You may use this tool to import data into Excel from a variety of different data sources, such as the databases that power the majority of accounting applications. More importantly, you might utilize Power Query to convert the data to become more useful to you. Such transformations could include removing unnecessary data columns, merging columns, sorting, inserting user-defined calculations, and filtering as a component of the query.

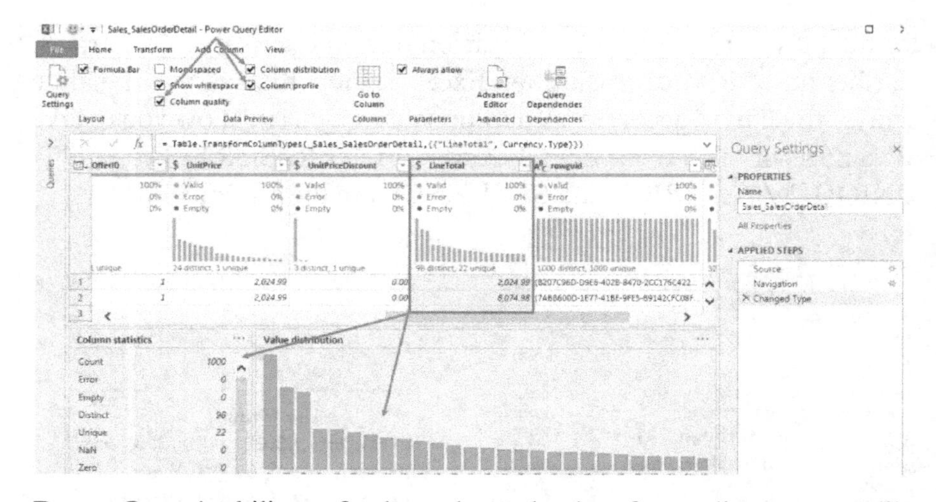

Power Query's ability to freely evaluate the data for quality concerns like completeness as well as accuracy is the latest enhancement. You can quickly identify possible problem areas with this function, such as incorrect data, incomplete records, and perhaps even duplicated records. Test the Column distribution, Column quality, & Column profile boxes upon its View tab on this Power Query Editor to take advantage of this function. Power Query creates the "quality snapshot" of and of the query's columns of data; however, tapping about any column within query displays a much more comprehensive overview of the data featuring statistics for such column and a graph of value distribution.

1.5 Tools

For the moment being, we've just listed a few of the many tools available in Microsoft Excel; in the following chapters, we'll go through them in greater detail.

SUM: This function adds all of the data values in the document's cells or a range.

COUNT: This function counts the number of cells in the statement that contain numerical data. This feature is helpful for quickly counting the number of items in a collection.

AVERAGE: This function of Excel determines the data's average values. It calculates a range's sum and divides it by the number of datasets in the collection.

MIN: The lowest value of cells involved in the selection is defined by this function.

MAX: The most significant value of cells involved in the selection is defined by this function.

1.6 File extensions or Types

.XLS

It is the most common file extension; this is the default extension for spreadsheets produced by Microsoft Office. This extension was widely used during MS Office versions before 2k7.

.XLSX

Spreadsheet files generated with Excel version 2k7 use this file extension. XLSX is the current and most recent default extension for such an Excel file.

.XLSM

This extension is created using an Excel 2k7 system. Excel macros are included in this category. With the aid of extension, you can quickly determine if a file contains macros.

.XLSB

If your excel files contain a large amount of data or information, this file extension will help you by first compressing, then saving, and opening them quickly.

.xltx

An Excel file is kept as a template that can use to create new Excel workbooks.

.xltm

An Excel file with macros that are stored as a template.

How can one figure out the extension within an excel file?

- Choose any file you would want to look up the extension for.
- Just right-click the File.
- Displays a drop-down menu of options.
- From the drop-down menu, choose Properties. Please see the image below for more information.

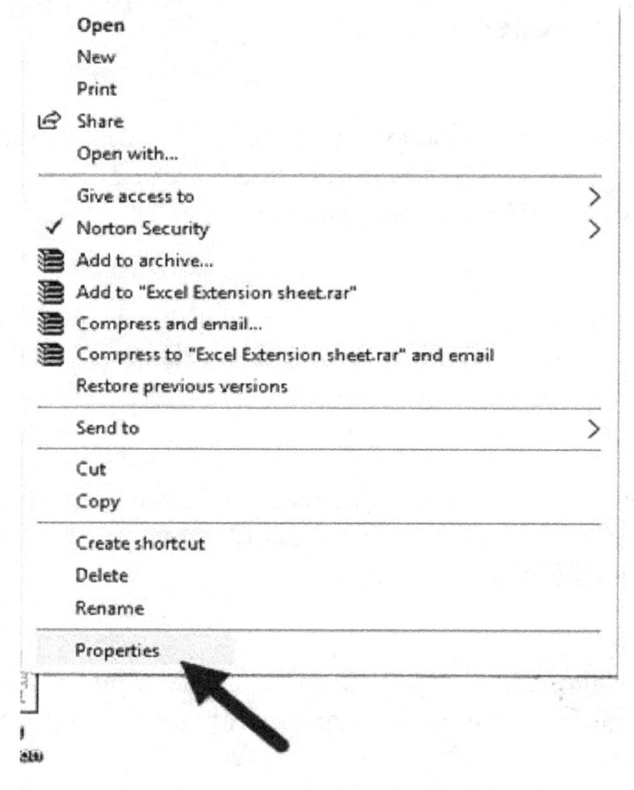

- A properties window will appear in front of you.
- File type /file extension is found in the file types section.

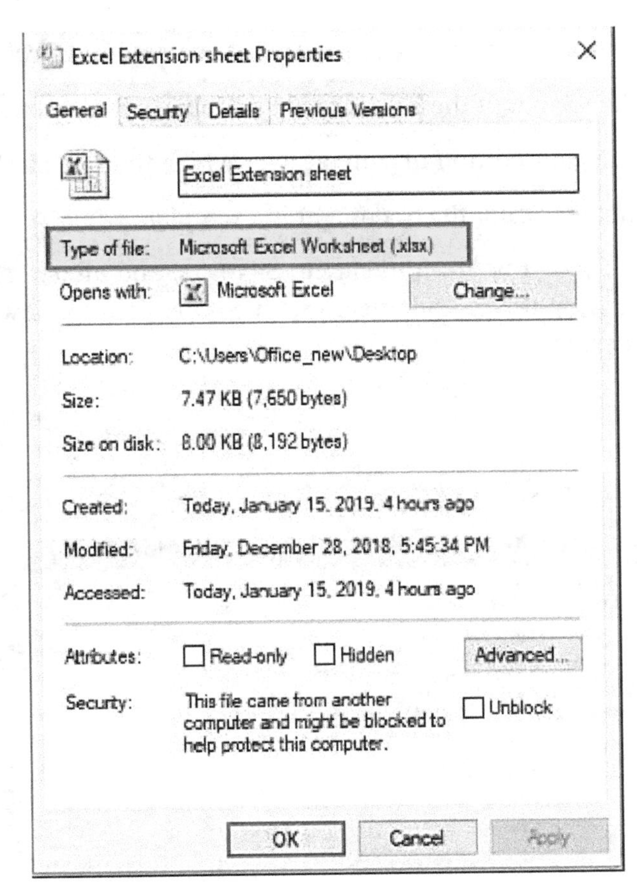

Note: Each file type may not support all of Excel's features, including specific formulas, and displays an alert message window. In this case, you'll need to modify the File's existing format to ensure compatibility.

1.7 The Importance of Excel File Extensions

Before users start an Excel file, the extension offers you valuable details about it. It also allows you to arrange the files which have been saved in the folders as macro-enabled files, template files, and other types of files. You can learn so much about an Excel file or how it's used just by looking at the file extension.

The extension of an Excel file tells you:

- If VBA or macros are used

- If you saved the File using an earlier version of Excel,
- If the format of the File is based on binary documents or XML
- The legacy edition in general with which the File has been saved
- Whether or not the document is a template.

In Excel, pick File from its menu, Save As, and afterward, choose that file type drop-down underneath its filename field to view most of your file extensions.

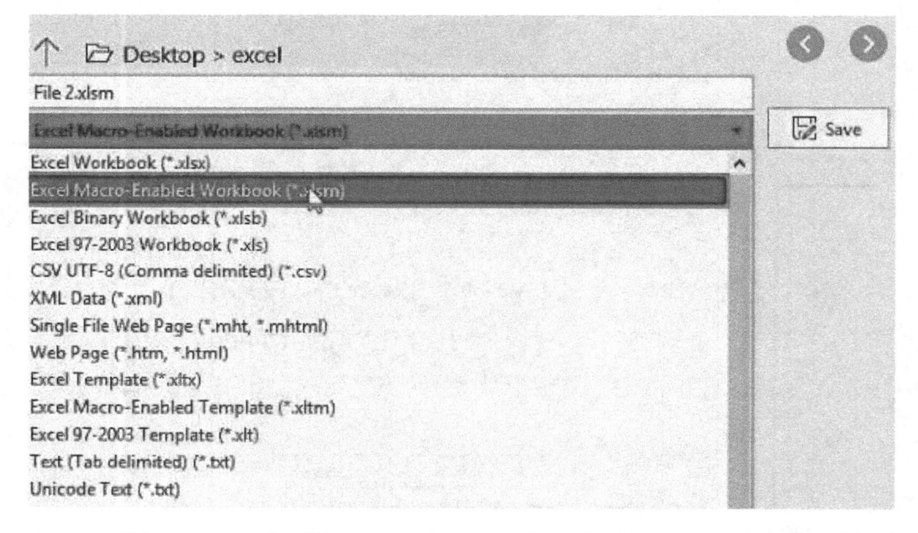

As you'll see, each file type has a descriptive name that aids in determining what the file extension means.

Chapter 2: Getting Started with Using Microsoft Excel

Microsoft Excel is software that allows you to create spreadsheets. People have to learn how to use Microsoft Excel as it helps them increase their productivity. Microsoft Excel is a simple-to-use program that is a crucial asset in any situation and critical for professional advancement.

2.1 From where to have Microsoft Excel?

Technology enthusiasts must appreciate Microsoft Office for countless times creating Microsoft Excel, which provides us with more headache relief than many other pain relievers. Microsoft's official website makes it simple to obtain this app. Use Google to find Microsoft's official website and click on the necessary links.

2.2 What is the right way to open Microsoft Excel?

To launch Microsoft Excel 2021, go over to the Start menu in Windows and pick Start All Programs. Then choose Microsoft Office, and finally, Microsoft Excel 2021. (If you don't have Excel, type it after clicking the start menu icon/selecting the tab.) A new, blank workbook has been launched, ready for you to fill in the details.

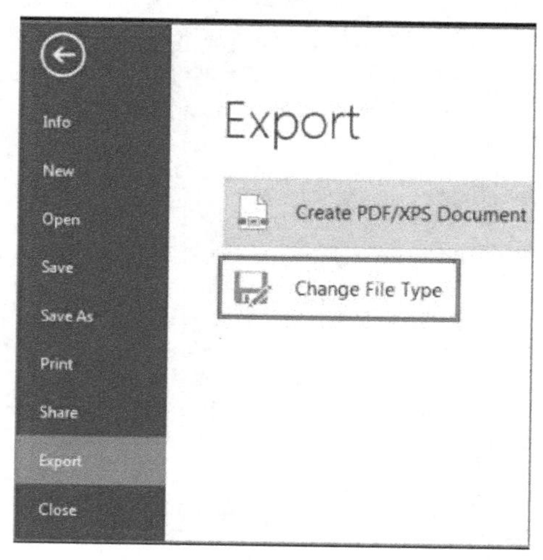

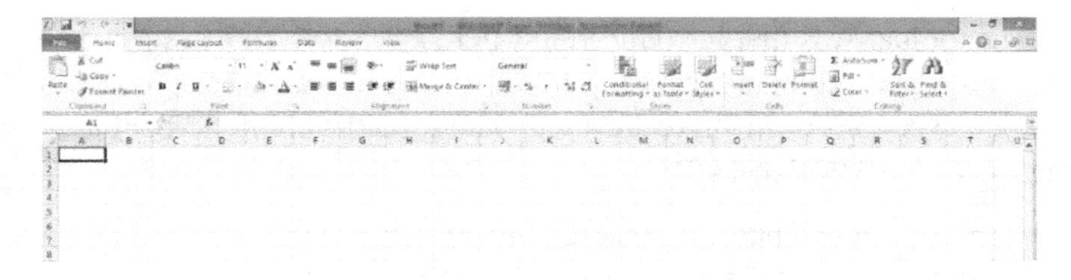

If you're attempting to open a file that has already been saved, you should first go through these steps.

Move 1: Open any XLSX file by double-clicking it.

To open XLSX files within Microsoft Excel, twice click the File. If you already have some version of MS Excel download and install on your computer (2016 or higher), double-clicking on a file will open it automatically.

Move 2: Open Excel and drag - and - drop files within it.

If you already have Microsoft Excel installed on your computer, you can drag the XLSX file into an open spreadsheet in Microsoft Excel with a single mouse click. To do so, pick the XLSX file, keep down the left mouse button, drag the File further into an Excel spreadsheet that has already been opened, and then release the mouse button. It will then enable the XLSX sort file.

Move 3: Choose "Open with" from the right-click menu.

The pop-up menu will also allow you to open the File if you have some version of Microsoft Excel installed on your computer. Right-click the XLSX file with your mouse and select the "Open with" option. Following that, a window will appear that will suggest programs to open the File in question. The said File would open when you click on MS Excel. If MS Excel isn't mentioned, you haven't installed that on your computer.

2.3 Excel ribbon and its components

The MS Excel toolbar seems to be a row of tabs and icons above the MS Excel window that allows you to easily scan for, identify, and use commands to complete a task. It appears to be a complex toolbar, and it is. The toolbar was first introduced in Microsoft Excel 2K7 to substitute the traditional toolbars but also pull-down menus that were present in previous versions of MS Excel. Microsoft added the ability to configure the toolbar in MS Excel 2010.

Sections, command buttons, groups, and launcher dialogues are the four main components of the MS Excel 2021 toolbar.

The toolbar tab contains many groups of commands that are logically separated.

The toolbar group gathers commands that are identical and are used as part of a larger task.

A dialogue box launcher would be a small arrow in the group's lower right corner that keeps bringing up more commands that are similar. A dialogue box launcher appears in groups of commands that are larger than the available space.

A Command button would be the one that you press to complete a specific action.

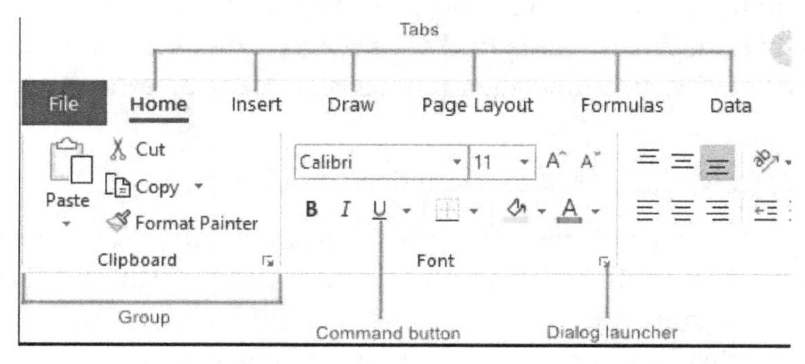

Standard Excel ribbon

File – Allows you to move to the backstage view, which contains the key file-related commands plus Microsoft Excel options. This tab was introduced in the MS Excel version of 2010 to substitute the Office button that was previously available in Microsoft Excel 2k7 version. In previous versions, it was recognized as a File menu option.

Home – This section contains the most frequently used commands, such as copy/paste, sort & filter, formatting, and so on.

Insert – is being used to add images, formulas, diagrams, PivotTables, headers/footers, hyperlinks, and special signs to a worksheet.

Draw – This feature is dependent on the computer you're using; it enables you to draw anything you want, whether with a digital pen and a mouse and just with the finger.

Page Layout – Provides tools for customizing the template of Microsoft Excel worksheets, along with onscreen tabs. Theme configurations, margins, grid lines, object orientation, page settings, & also print fields are all monitored by these tools.

Formulas – This section contains tools for incorporating features, naming them, and handling calculation options.

Data – Hold down the complete command to handle the data in a Microsoft Excel worksheet and link it to external data.

Review – It allows you to fix spelling errors, make tracking improvements, add feedback as well as notes, and save Microsoft Excel workbooks & worksheets.

View – Move between worksheet views, view, freeze panes and organize different windows.

Help – This tab provides quick access to the Help Task Pane of Excel, allowing one to notify Microsoft Support, request feedback, suggest new features, and quickly access training videos.

Developer's mode – gives one access to more advanced features such as Visual Standard Application macros, Microsoft Form controls plus Microsoft ActiveX, or even XML file commands. Since this tab is hidden, you must first toggle it before using it.

Add-ins – When you want to open some old workbook and sometimes activate an add-in that customizes the menu and a toolbar, this setting appears.

2.4 Tailor the work environment by understanding your worksheets

When you first start Microsoft Excel (by double-clicking its icon or choosing it off the Start menu), the software would ask you what you want to do. If you want to start a new spreadsheet, select the Blank Workbook option. A fresh workbook, including one blank sheet, will be created by Microsoft Excel.

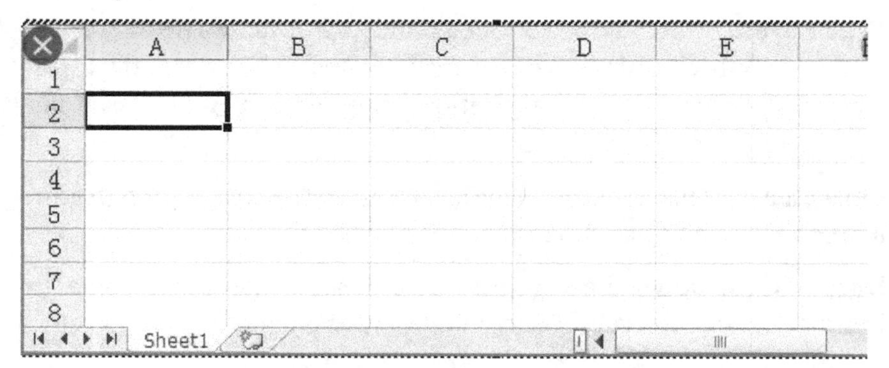

The horizontal axis has been interpreted as rows, whereas the vertical axis has been classified as columns; similarly, a selected box would be called a cell.

Customizing the Excel Workplace

The Microsoft Excel framework is designed to reflect things like using an excel application or how people use the MS Excel software. The components of Microsoft Excel are being described below. The Microsoft Excel GUI aims to make MS Excel workbook operations as well as procedures more effective.

The Live Preview feature shows how the frame's formatting has improved. To see the format in the browser, drag the mouse cursor over the command.

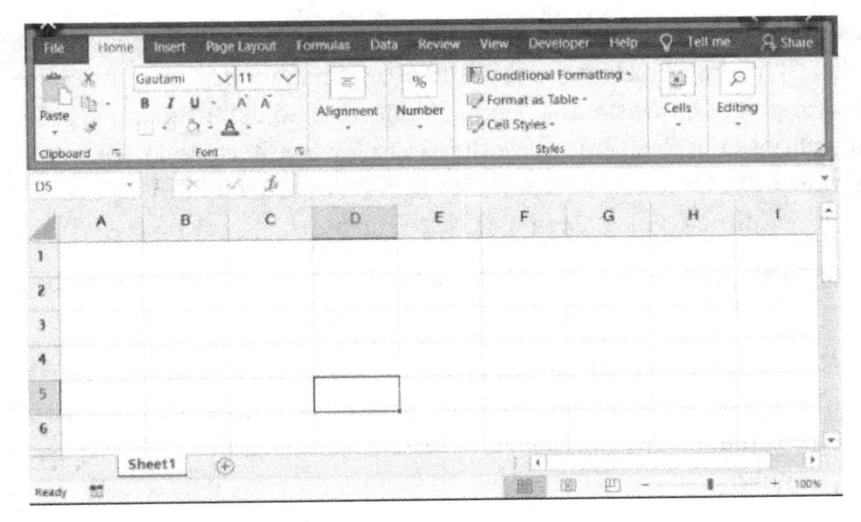

A few of the following configurations can be used by users to customize the working GUI.

Making a template in Excel:

Open Microsoft Excel and create a new blank workbook.

Workbook must be saved with a specific file name in the specified folder.

Below are some additional tips and more detailed measures.

One must alter a few Excel workbook basics.

For example, font style and size: highlight the parts from each worksheet that you want to change, and then choose your options for number, alignment, and font style from its Font category just at the top of that same worksheet.

Column size/layout: Typically, you choose different column widths, select the columns or even the entire working sheet, & afterward change the width of the selected column.

In Print Settings, choose one or perhaps more worksheets, then go to Page Layout and then to Page Setting group so as to customize print settings including header or footer, page orientation and also page margins, as well as many other print layout settings.

Gridlines: Would you choose the gridlines on each worksheet to be darker? The dark borders or grid lines are visible but do not print. Select File> to go to Options > and then Advanced to adjust the gridline color. Then choose Display options for the existing worksheet, and then choose the title of its workbook from the drop-down menu. Finally, pick a specific gridline color under Display gridlines.

Worksheet count: one can insert and delete worksheets, as well as rename sheet tabs & adjust the color of worksheet tabs.

NOTE: When you introduce a new worksheet to your customized default workbook, the actual layout & formatting will be restored. You'll choose to attach extra worksheets to just the actual workbook to set aside an optional or main worksheet which you might copy if necessary.

2.5 Ribbon Customization

In Design, a Ribbon window is where you do the majority of your MS Excel Ribbon customizations. It's an option in Microsoft Excel. So, if you're going to start customizing the ribbon, you'll have to get one of the following:

Pick the Customize Ribbon option from the File tab > in the Options command

Right-click to customize the ribbon... from the context menu:

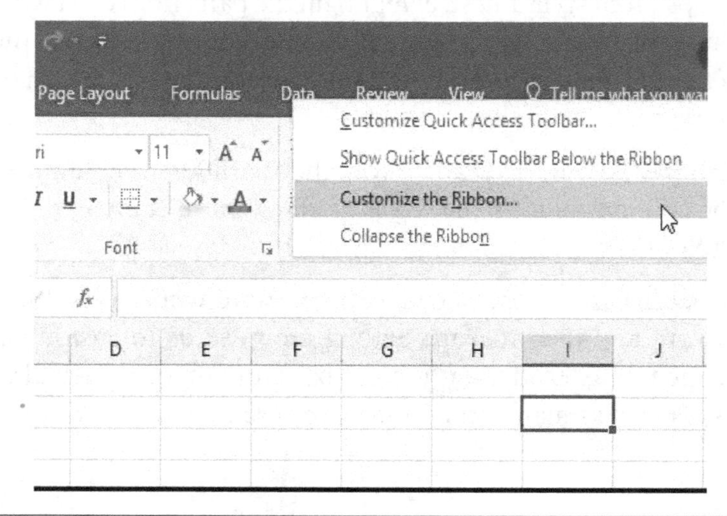

Introduce additional tabs in a ribbon

You can add your tab to that same Microsoft Excel ribbon to make your favorite commands easily available. Here's how to do it:

Choose the bottom list in tabs in the Customize Setting and Ribbon window, then press the New Tab Logo.

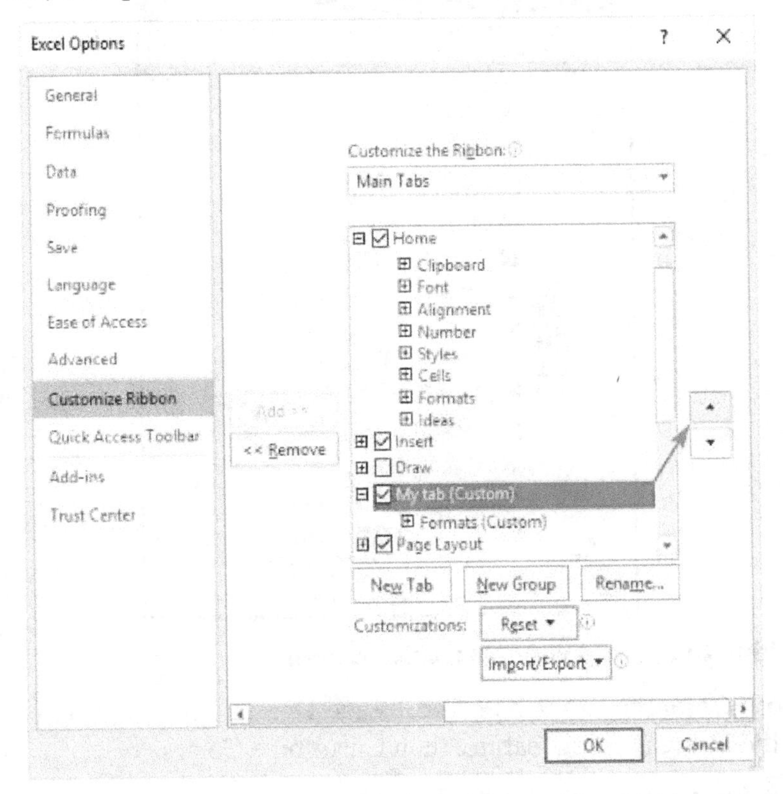

Since this would add a custom tab to a custom group, commands can only be used to add to one of the custom groups.

Select the Fresh Tab (Custom) tab you just created and click the Rename button to properly name. Similarly, change the default name given by Microsoft Excel to a custom category with more detailed instructions. When it gets done, click OK to save the changes.

2.6 Deciding the theme's colors

Select Colors from the On-Page Layout Configuration tab within Microsoft Excel and pick a color you like.

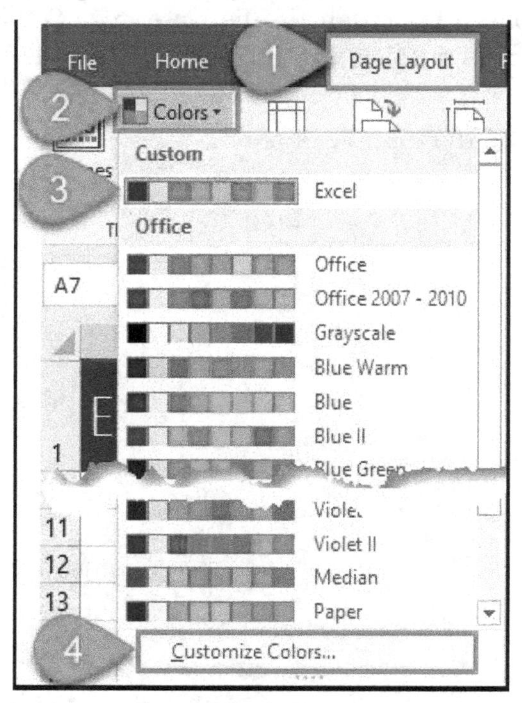

Working to develop your own color scheme

Choose Colors from the MS Excel Page Design configuration tab and also the Design setting feature, then Customize Colors.

Select a good color for Colors' theme by pressing the button that is next to the theme color you want to change (– for example, Accent 1 & Hyperlink).

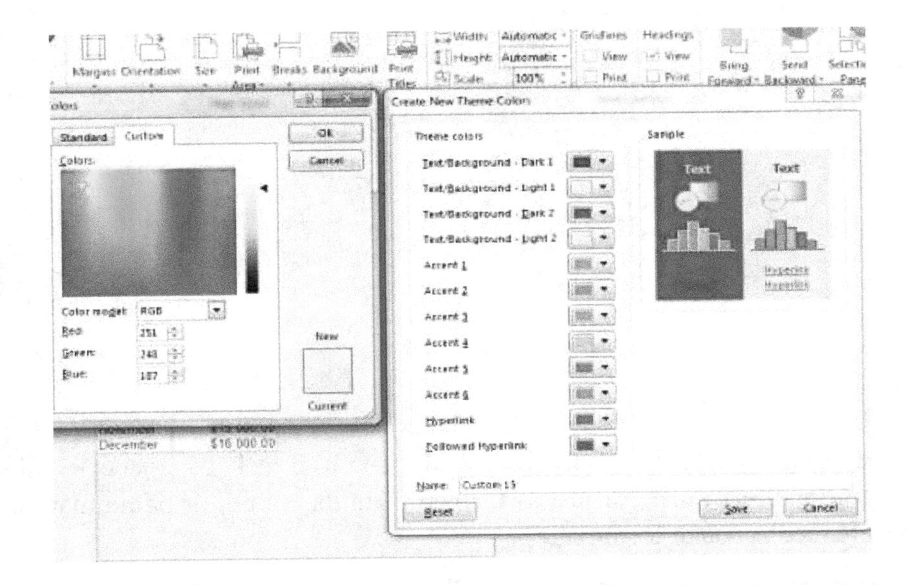

Go With more Colors, pick a good color from the Regular tab, and then insert code numbers of color, or have a color from its Customized tab to create your color theme.

A sample pane is where you can see a preview of the adjustments you've made.

You could repeat this procedure for the colors you want to alter.

In the Name Box, type the name you need for the new color scheme, then press the Save button.

2.7 Setting the formulas

First, go over towards the Formulas icon> Referred as Calculation group on a Microsoft Excel toolbar, then press the Calculation Options button and select the below options.

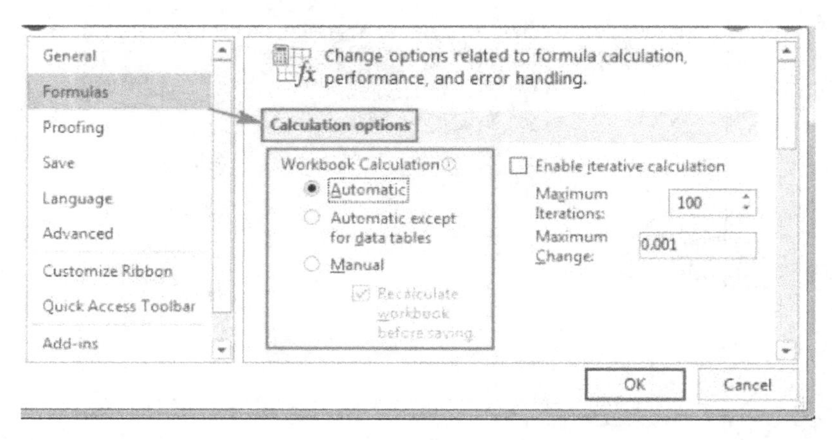

Automatic (default) - Instructs Microsoft Excel to recheck all conditional formulas for just any given formula, value, or name that is referenced for such formulas.

Automatic Excluding these Data Tables - Verify all related formulas automatically, except for those of the data tables.

Remember to distinguish between Microsoft Excel Tables (that is, Insert > Table) and Data Tables (Data > then What-If Study > and then Data Table), which also approximate numerous formula values. This choice disables automatic recalculation for data tables only, whereas regular Microsoft Excel tables are still calculated automatically.

Manual - disables Microsoft Excel's automatic calculation. Rechecking open workbooks is only possible if you're using any of these approaches.

2.8 Proofing options

Proof configurations change how Microsoft Excel corrects and then format text as you write – this feature allows you to choose the configurations that are being used to automatically correct text as you type, as well as to save and reuse text or other items that you use frequently. Go to AutoCorrect Options, then choose the option that you want.

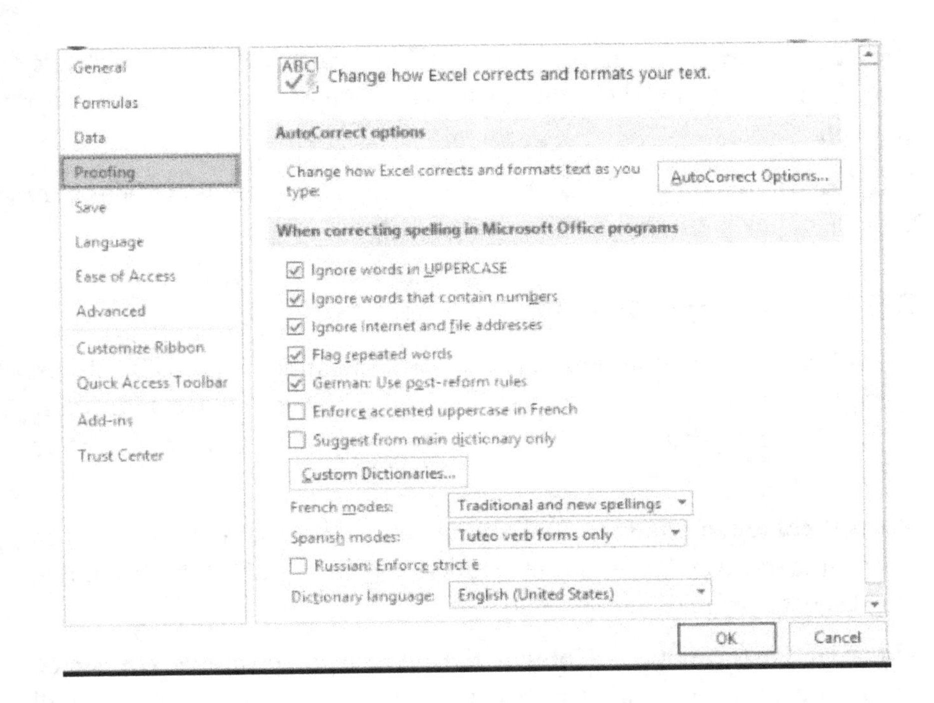

Words in UPPERCASE are ignored: In Microsoft Excel, In a spelling search, words wrote in UPPERCASE characters should be ignored.

Ignoring words with numbers in a spelling search: Microsoft Excel-Ignore particular words with numbers during a spelling check.

Neglecting Internet and file addresses: If one wants Microsoft Word to automatically overlook Internet addresses, file names, as well as e-mail addresses, click this check box.

Flag repeated words: During every spelling search, this feature detects and marks words that are repeated.

Impose the accented uppercase: It displays magnified uppercase characters for French material.

Custom Dictionaries: Use this button to choose the vocabulary you want to use when testing spelling.

French modes: Sets spelling guidelines also for pronunciation for French words in the dictionary. In the above list box, pick the option you would like to keep.

Spanish modes: You must create spelling rules for Spanish words. Choose a separate option from the drop-down menu that will not appear in the list box.

Dictionary language: It enables users to choose which dictionary language they want to use.

2.9 The best tips for working with Excel

A collection of standard criteria for creating Microsoft Excel spreadsheets can be found here. They will help you in working more efficiently and produce more structured data, so you must adhere to them.

Worksheet Development - By keeping all pertinent data in one tab, you were able to use Pivot Tables, functions, also Subtotals, including Worksheet Formulas.

Updated Performance - Compared to a smaller group of connected workbooks, fewer, bigger workbooks have improved overall performance.

Data Layout - Viewers normally search the rows & columns to get a sense of how the data is organized. Viewers can more effectively manage and evaluate data by first defining and presenting the most accurate information.

Protect the Cells - You can protect private cells and delicate and ranges by limiting which users are allowed to edit or format them.

Data Validation Function - Any system that can assist in the removal of errors is definitely a time-saving feature that can keep the data up to date. This function can be found therein in the Data tab of the Data Tools category.

The Benefits of Using Color -Color is a great way to illustrate important details and give readers a break when reading a lot of information. It would be best if you used the various color options available in Conditional Formatting and Cell Types, as well as the standard color options.

Absolute References - Do you need to keep the same cell's reference when copying or while using AutoFill? To avoid automatic content changes, use the "$" symbol.

Make a note for Simple calculations- When exchanging and creating formulas, mark the ranges and give the formulas a descriptive name. This will make it easier to select large amounts of data and comprehend the different formulas' intent.

Summary Sheets -Design the named range to just group the totals from every sheet you want to create a summary. If you'd like to summarize the total sheet, go to the feature menu and select the named range.

Using the Cell Merge Alternative- To make sorting easier, go to Format Cells, select Orientation, and then use Center By Range from the horizontal drop-down to center a label through several cells.

Chapter 3: Standard Formulas and Essential Calculations

Each formula, and even some Microsoft Excel functions, are crucial. You will find it more meaningful and valuable in both cases, and you will love doing that in a spreadsheet. This chapter explains the fundamental concepts you'll need to know to use these formulas with confidence in Microsoft Excel.

3.1 What is the best way to enlist formulas in Excel?

For calculations, Microsoft Excel uses the regular operators, such as a plus mark (+), a minus mark for negation (-), an asterisk symbol for multiplication (*), a front slashing symbol for division (/), and a caret () for exponents. The key point is that in Microsoft Excel formulas, you must commence them by an equal (=) symbol. Since the cell contains, or is equal to, the formula but also to its value, so this becomes important.

In Excel, here's an instance of how to make a basic formula:

- Select the cell where the answer emerges (B4, shown example).

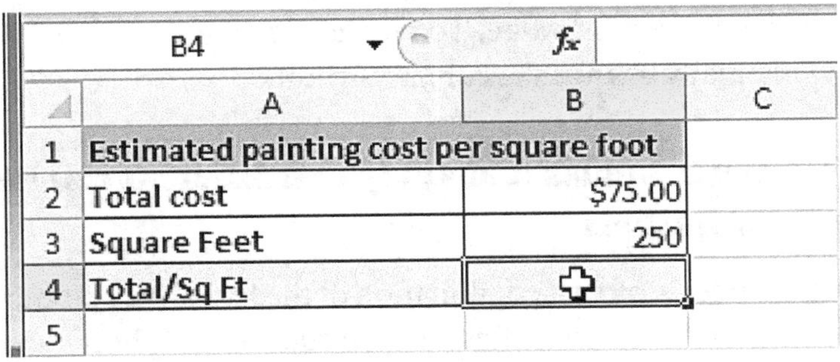

B4 fx

	A	B	C
1	Estimated painting cost per square foot		
2	Total cost	$75.00	
3	Square Feet	250	
4	Total/Sq Ft		
5			

- Substitute a symbol for the equal sign (=).
- Fill in the formula you want Microsoft Excel to quantify (for example, 75/250).

MAX ▾ × ✓ fx =75/250

	A	B	C
1	Estimated painting cost per square foot		
2	Total cost	$75.00	
3	Square Feet	250	
4	Total/Sq Ft	=75/250	
5			

- The calculation will be completed by the formula by hitting an Enter key, and now the result will also be displayed in the specified cell.

B4 ▾ fx =75/250

	A	B	C
1	Estimated painting cost per square foot		
2	Total cost	$75.00	
3	Square Feet	250	
4	Total/Sq Ft	$0.30	
5			

- If the figures in a formula result are high in a cell and appear as a (#) hash rather than just a value, it means the columns aren't big enough to represent the cells' value. To see the contents of its cells, you must now manually change the column's width.

3.2 Creating formulas that apply to a different cell inside of the same worksheet

Each cell address is often a combination of the letter of a column and the number of the row that classifies a cell in such a worksheet.

The mentioned points shall be considered when creating cell references on a similar sheet.

To begin, select the cell into which you want to paste the formula. Then type the equal sign (=).

Now take the following steps:

First, you must type the correct reference in the appropriate cell or perhaps on the top bar known as the formula bar.

Choose the correct cell in which one you want to refer.

Finish by typing the remaining formula then pressing Enter.

For example, add the values in cells A1 & A2, type this = symbol, click A1, type this + logo, press A2, and finally press Enter.

	A	B	C
1	5		=A1+A2
2	10		

Click the cell to make a cell reference

In Microsoft Excel, identify the suitable cells on a particular worksheet if you'd like to create a range comparison.

For example, to add the separate values in cells A1and also A2 or A3, type an equivalent symbol, then the title of the SUM method and thus the starting parenthesis, then pick the cells from A1to A3, type the ending quote marks, and then click Enter.

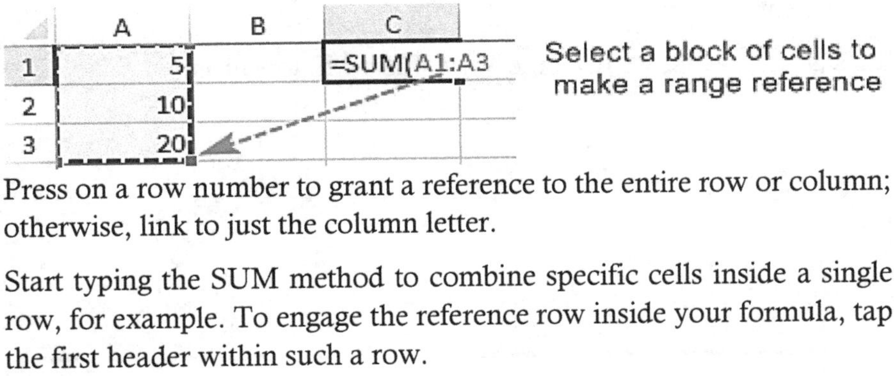

	A	B	C
1	5		=SUM(A1:A3
2	10		
3	20		

Select a block of cells to make a range reference

Press on a row number to grant a reference to the entire row or column; otherwise, link to just the column letter.

Start typing the SUM method to combine specific cells inside a single row, for example. To engage the reference row inside your formula, tap the first header within such a row.

3.3 Is it possible to build a formula that connects to other workbooks?

To prevent duplicate data in separate sheets, use Making Links and similar external cell references. This saves time, cuts down on errors, and improves data integrity. To create an external connection, take the following steps:

Open every one of the workbooks first.

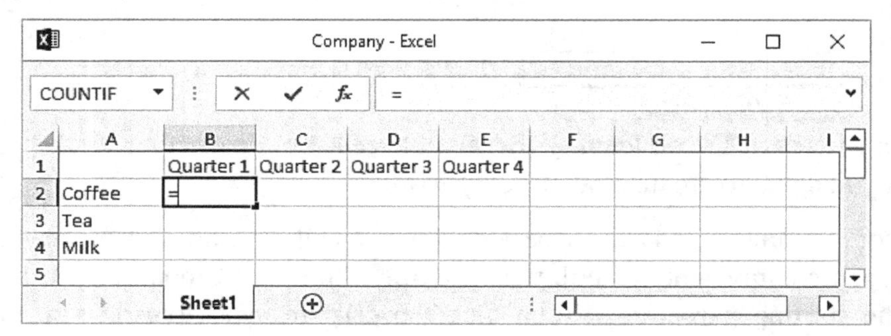

Go to the "Company" Workbook and select cell B2, after which type an equivalent symbol =

Pick Switch windows from the View Tab's window group.

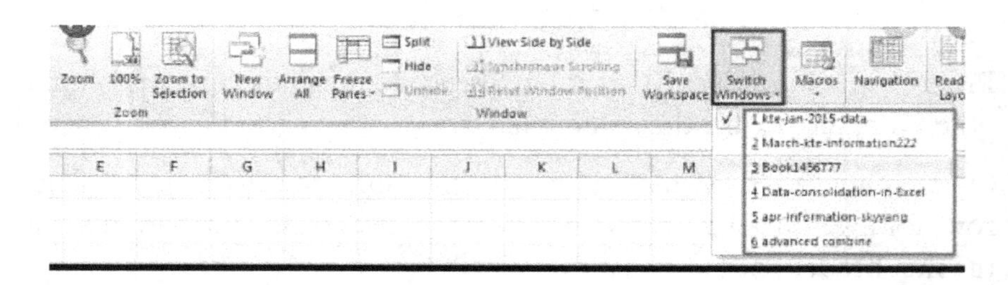

Next, choose the "North" workbook.

Choose cell B2 in the above workbook.

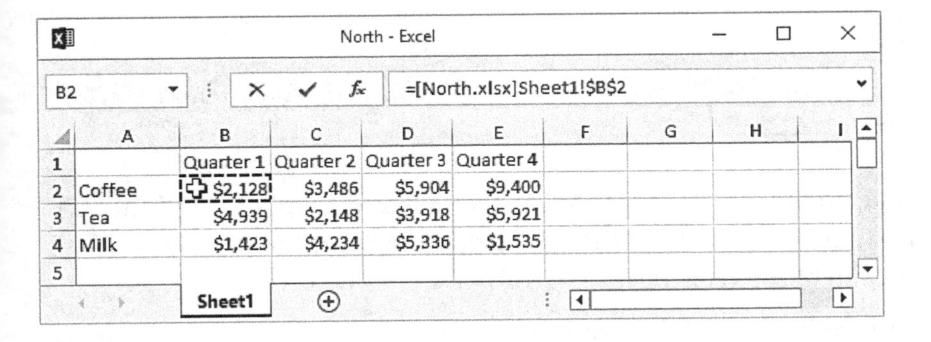

Next, type a + and then replicate steps 3–6 for the "Mid" Workbook.

Now, for the South workbook, replicate steps between 3 to 5.

Finally, remove the $ signs from the cell B2 formula before continuing.

Results:

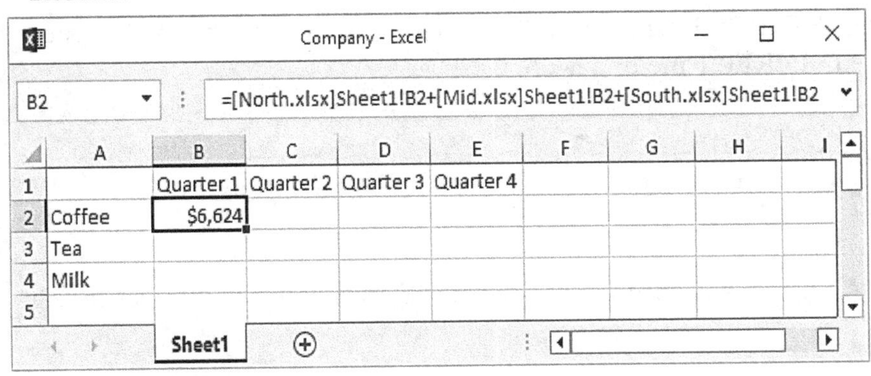

3.4 How do you use apostrophes to enclose and complete the file and worksheet names?

You may modify the text string for proper reference by using Excel's INDIRECT form. The & operator can be used to create a text string.

Have a glance at the INDIRECT form, which is described below.

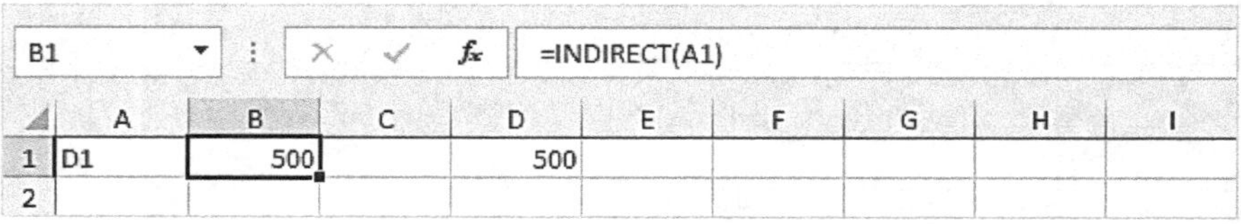

B1 | fx =INDIRECT(A1)

	A	B	C	D	E	F	G	H	I
1	D1	500		500					
2									

Explanation: =INDIRECT(A1) is simplified to (symbol) =INDIRECT ("D1"), however the INDIRECT method replaces that text string "D1" with a right cell reference, so =INDIRECT("D1") becomes =D1.

The simple INDIRECT feature, as shown below, produces a nearly identical result.

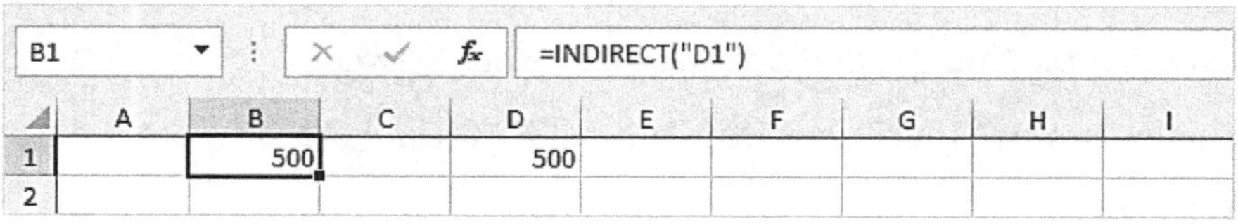

B1 | fx =INDIRECT("D1")

	A	B	C	D	E	F	G	H	I
1		500		500					
2									

Is the INDIRECT function still required? Sure, this is what will happen if the INDIRECT method is not used.

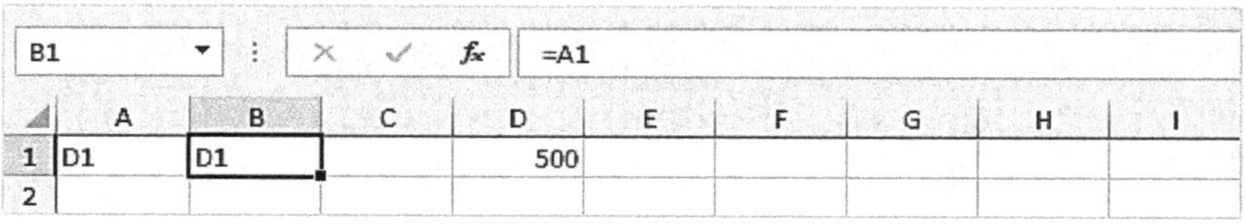

B1 | fx =A1

	A	B	C	D	E	F	G	H	I
1	D1	D1		500					
2									

The & operator is used to connect the "D" string to the number in cell A1.

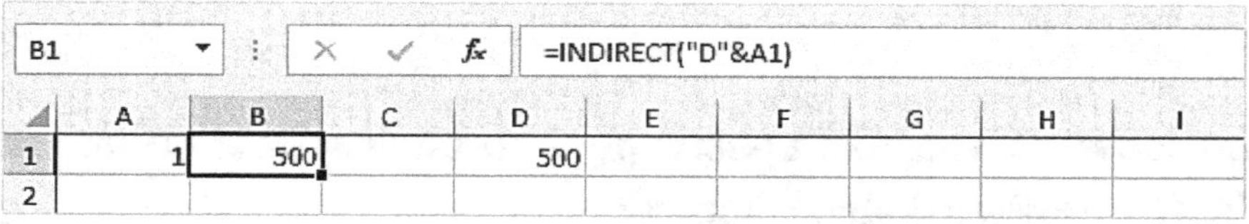

B1 | fx =INDIRECT("D"&A1)

	A	B	C	D	E	F	G	H	I
1	1	500		500					
2									

Explanation: the above formula is abbreviated as =INDIRECT ("D1"). =INDIRECT("D1") gets simplified to =D1 once more.

Chapter 4: Excel Workbook (Choose, Edit Or Select Cells)

The most popular functions used in a Microsoft Excel worksheet are triggering, selecting, and changing cell values. Anyone that uses Microsoft Excel must've been able to activate a cell and keep it active when inserting data into this one. To obtain an enabled cell address, one must first allow it. Even so, not everyone understands what an activated cell is. So here's where we'll begin:

4.1 What does it mean to have an active cell, and how do you go about activating one?

In Microsoft Excel, an active cell seems to be usually a rectangular cube that displays the cells. This makes it easier to see which cell we're working on or where the data was entered. The current Cell, cell pointer, or selected Cell are all terms used to describe the active cell.

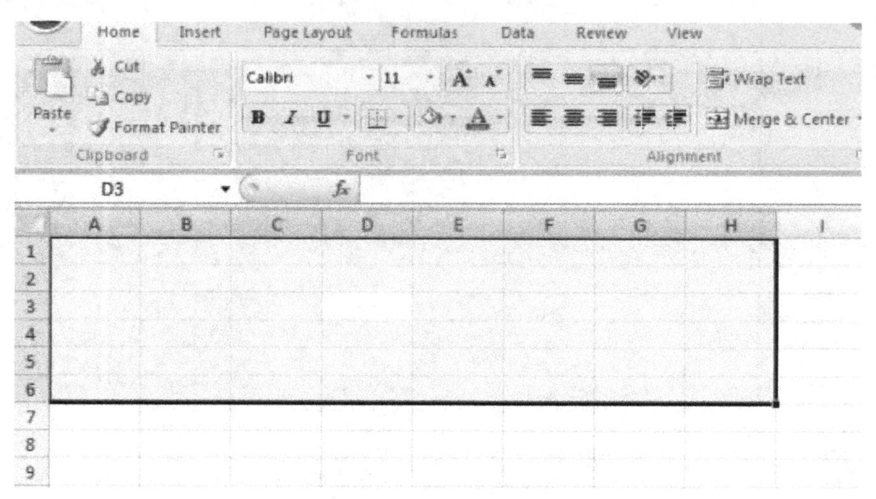

Every user of Microsoft Excel can say the distinction between an active cell and one that is in edit mode. It's also crucial to understand how we activated a cell.

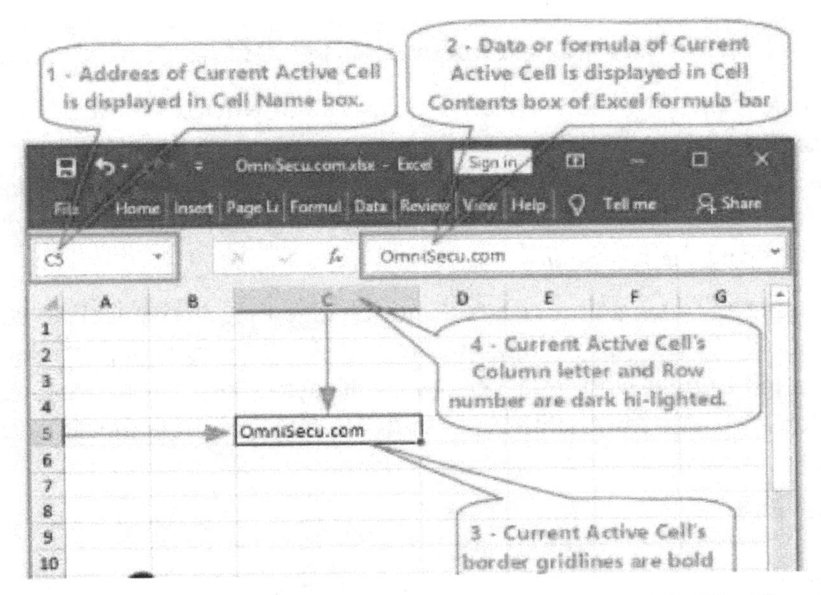

When you click on a cell, it means you're going to unlock it. You'll notice a grey and green line running across it. The term "active cell" refers to a cell that is ready to be moved to edit mode. In this quick example, we'll learn how to bring a cell into edit mode.

By using the Down arrow Key, Left arrow, Up arrow, and also Right arrow Keys, we can adjust the active cell by dragging the mouse cursor & selecting another cell. The row of active cell would move down as soon as one presses the Enter key, so if you make some adjustments to that cell, you would have to manually select it again before putting this in editing mode.

4.2 How do you put an excel cell into "editing mode?"

Consider pressing the "F2" key for any cell of Microsoft Excel. In that case, the present model would be changed to "Edit," and the Cell mode would be changed to "Editing" when you twice click on a non-blank cell of (the one in which data already has been actually entered) Microsoft Excel using the mouse pointer.

Tapping the navigation keys will not change that Active Cell to some other cell if you're in "Editing" mode of a cell. The text cursor can change from left to right within the cell toward its arrow key if you switch right or left with its arrow buttons in "Editing" mode, as well as the up / down arrows have no effect in "Editing" mode.

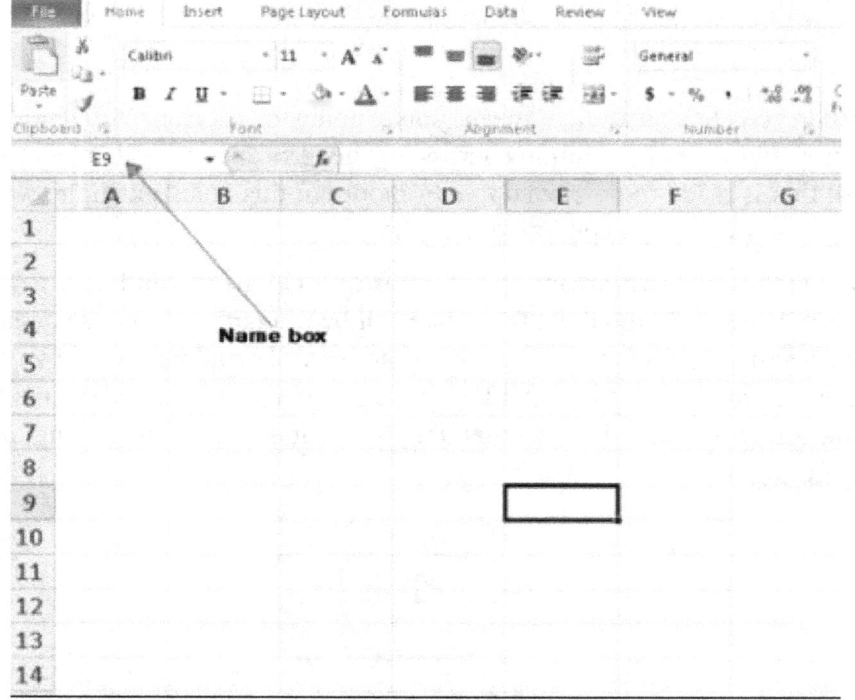

With the "F2" function key, one can switch from "Enter" to "Editing" mode while being in "Enter" mode. By hitting the "F2" key, one may easily switch across "Enter" & "Editing" mode. Pressing any Arrow key when in "Enter" mode will cause the current cell to lose focus, like in an Active Cell. The Active Cell will then be the next cell in the line of its Arrow key.

4.3 What is the best way to keep a cell inactive form?

Excel 2013-2019 editions cover the active cell around the dark border, so identify the cell you're editing. Once the cell has become an active cell, press the Enter key to advance to the next cell. If you're assessing the effect of changes to the original cell value, this new setting encourages you to take a step back before you modify the value again. Nonetheless, Microsoft Excel prevents you from moving the cell, and it remains active unless you manually change its location.

Select "Options" from the "File" tab.

Choose "Advanced" from the left-hand menu within Microsoft Excel Options.

In the Options of Edit section, uncheck "Shift Selection After Clicking Enter."

Click the "OK" button.

4.4 Getting back to an active cell?

Some spreadsheets are huge, extending well beyond the top and bottom of the computer screen. There are many ways to maneuver through a large spreadsheet, but it's common to go up and down than lose focus from its active cell. A typical scroll down or up will take you back towards the active cell, where you were before. Returning to an active partition on a large sheet is extremely difficult. Most of the time, you often have no idea about the directions to the active cell; however, it refers to how long you've been wandering. The least organized way to get back to its active cell is to scroll backward.

If you recall the active cell, press [Ctrl]+G, type in the cell address & range name, and afterward click OK. You'll almost certainly miss the cell reference, but that's fine since any keyboard shortcut key can guide you back to that same active cell—from anywhere on the very same sheet. After you've lost track of the active cell, press [Ctrl] and [Backspace], & Microsoft Excel would return you to just the active cell.

Chapter 5: Usage Of Find and Replace Features

Microsoft Excel's Method to Replace and Find feature is used to locate items in the workbook, such as a text string or specific quantity. You may also discover the search entity for future reference and replace it with another. In your search words, you can also include wildcard icons, including question marks & asterisks, as well as numbers. You can quickly scan columns and rows, inside comments as well as values, and within the worksheet or the entire workbook. One may promptly check columns and rows and also search within comments.

5.1 Finding information with the Find function

To begin, select the complete range of cells for search within the workbook. Pick any cell on the currently active sheet to scan the entire worksheet.

By pressing the Ctrl & F keys at the same time, open Excel and select the Replace & Find mode.

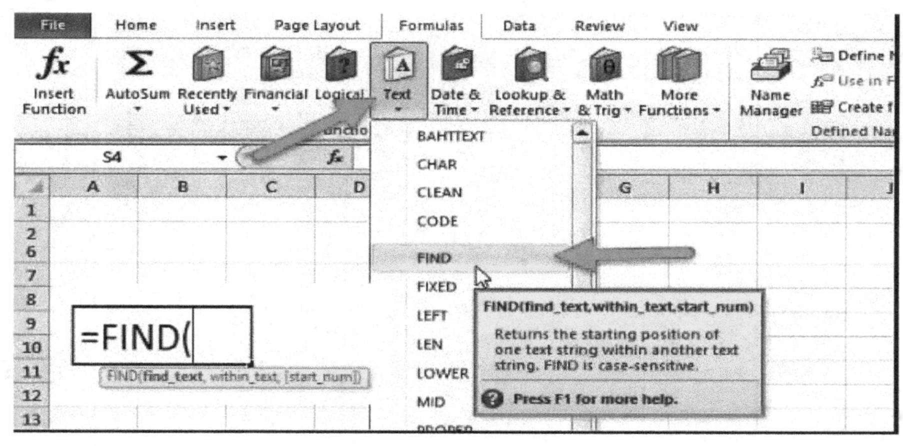

Type the characters (numbers or words) into the find box, and then click "find all" if you're looking for all observations; otherwise, click "find next."

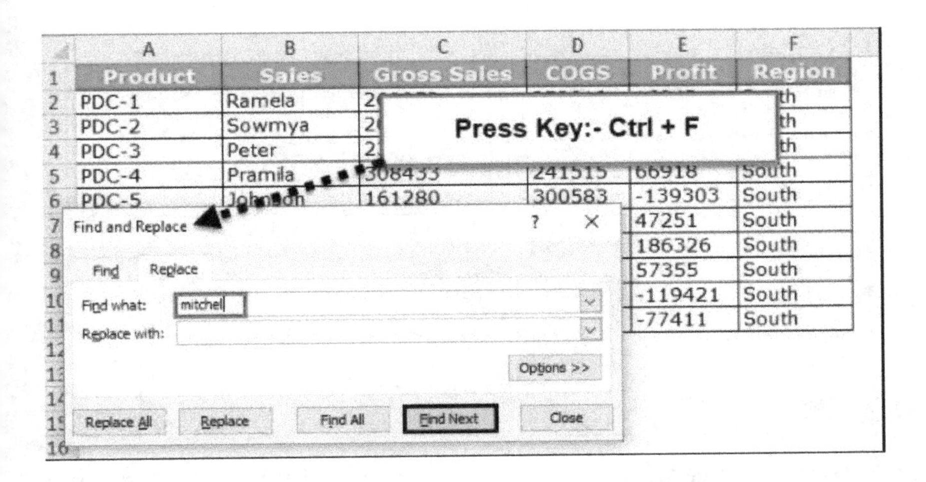

When you click Find Next, MS Excel will choose an existing word in the document that matches the quest. The following press will pick the second word that already exists, and so on.

5.2 Using Excel's Replace Feature

You can use Microsoft's Replacement option to replace any Microsoft Excel symbols, which could be text and numbers. The measures are outlined in detail below.

Select a set of cells where the text and numbers should be replaced. You can click in any cell on a presently active sheet to substitute the character(s) for the entire worksheet. Ctrl + H is indeed a keyboard shortcut command which might assist you in launching the replace and find window.

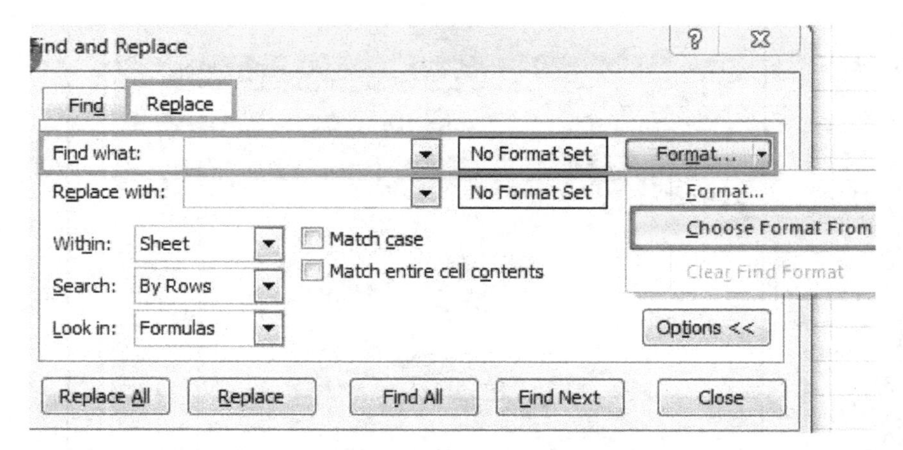

Alternatively, go over to Home > Editing group > find & Select > now replace in the Editing group.

Wildcards have been used to find and replace features.

One may conveniently automate numerous replace & finds tasks in MS Excel by using wildcard characters while your search:

To find a string of characters, use the asterisk (*) wildcard. For example, sm* will find both "smile" as well as "smell." To find a specific character, use the doubt mark (?). Gray, e.g., will display "Gray" or "Grey."

Let's say you would like to seek a list of terms that start with "ad," and the search criteria is "ad*." Here's something to keep in mind: if you leave Microsoft Excel's default options alone, it will scan only for the requirements anywhere in a cell. It will help all of these cells that have "ad" in some form in other situations. To prevent a situation like this, go to the Options icon and look for the Match entire cell content tab. As seen in the screenshot, this will cause Microsoft Excel to only retrieve values that begin with the letter "ad."

You can enter the tilde (-) mark before the authentic asterisks/query marks in the Microsoft Excel worksheet whenever you'd like to find them. E.g., if you wanted to find cells with asterisks, you'd type -* in to Find What box. Using to locate cells that contain question marks? In accordance with the search criteria.

That's how one can quickly and easily substitute complete question marks (?) within the worksheet by one or more values (in this case, number one).

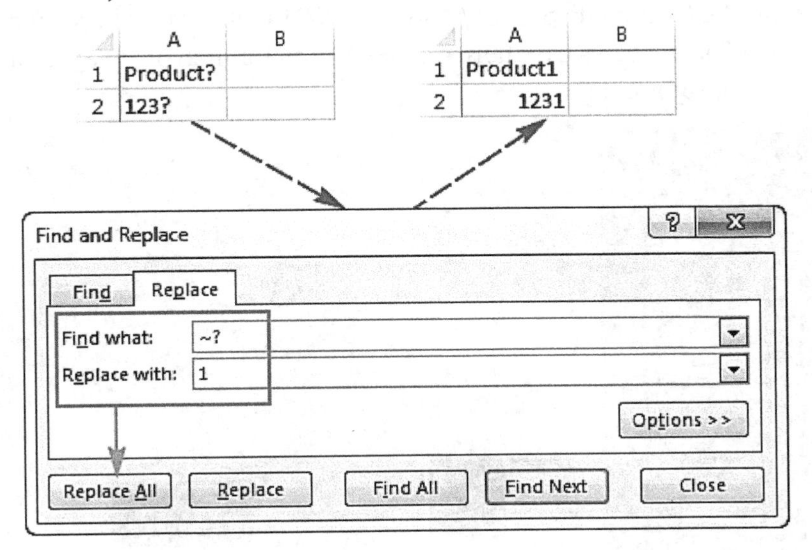

Chapter 6: All about the Main Screen

When you enable MS-Excel, the first screen you see is the beginning screen. The Launch display of MS-Excel 2021 is shown in the screenshot shown. There are two panels on the Excel Launch Display:1) a left panel, 2) a right panel,

Several newly accessed Excel data can be found in the green panel on this left side. That "Open another workbook" portion includes a list of recently opened MS Excel sheets. When you select "Open Other Workbooks," MS-Excel displays the backstage view, which allows you

to navigate to a new Excel folder. "Blank workbook" is an essential click upon this right panel. By selecting the "Blank workbook" button, a new blank workbook of MS-Excel will be created. As shown in the image below, the most recent Excel workbook contains a single initial worksheet (labeled Sheet-one) in the updated workbook (labeled Book1).

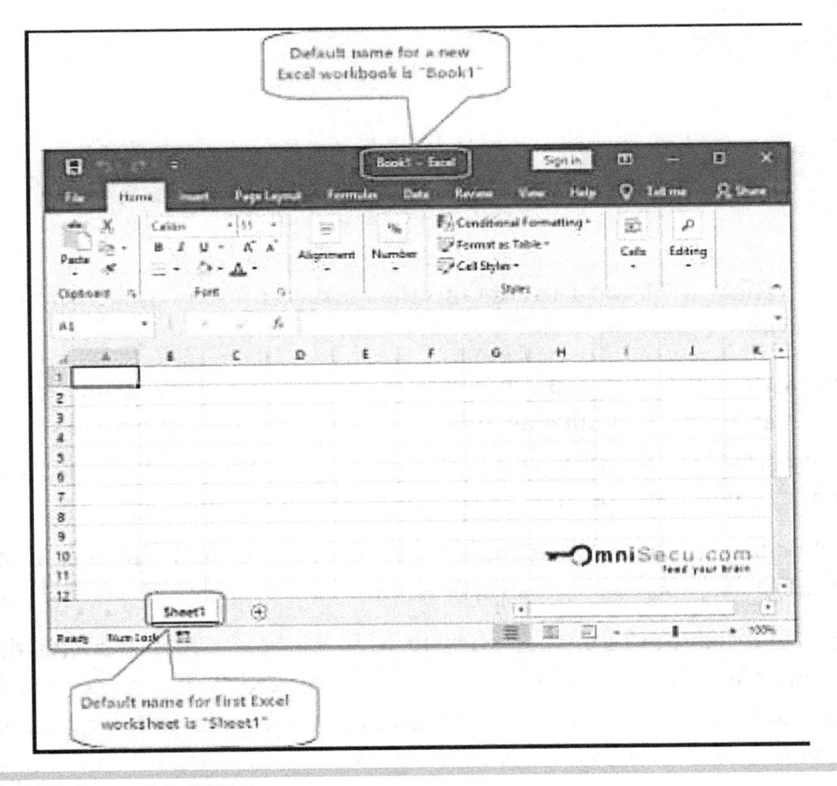

6.1 Using the Ribbon

The Excel Ribbon has been the first topic covered in our MS-Excel Introduction. The ribbon is a key component of the MS-Excel app, providing links to many of its features, facilities, and amenities. It was first announced with MS 2007 Office. This is no longer the trend for all sophisticated product packages. There are a variety of other options for controlling Excel functions and avoiding that MS Ribbon.

Home Tab

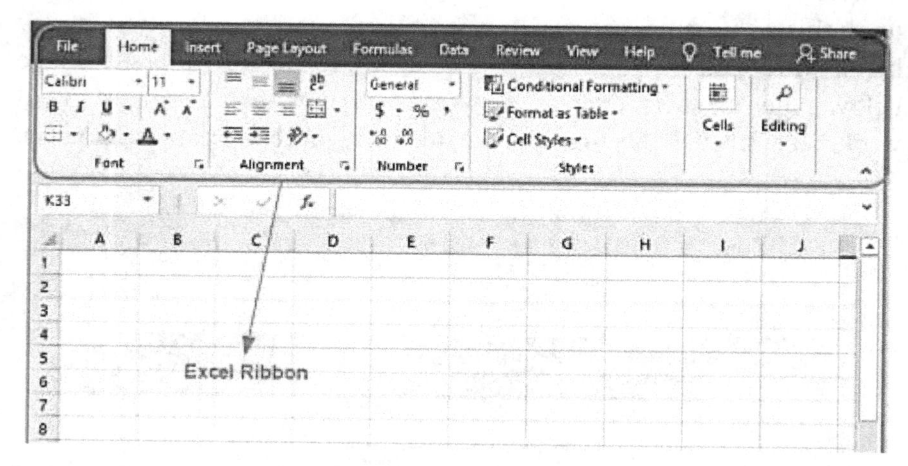

In MS Ribbon, Home Tab contains the most commonly used formatting and editing choices.

The Clipboard offers a convoluted method of performing simple clipboard operations such as Split, Copy, & Paste. (Most people tend to be using keyboard shortcuts keys (Ctrl-V, Ctrl-X, and Ctrl-C) or the right-click menu to pick specific options.)

Font allows you to customize the layout of text in the Excel spreadsheets, such as font, size, & color.

The Font has the Launcher press in MS- Ribbon, which displays that Format Cells box & also the Font tab is activated. We can merge cells and track both vertically and horizontally orientations; the orientation provides text flipping & wrapping commands—useful for longer headings.

An Alignment function in Excel Ribbon involves a launcher icon that displays its Format Cells' box, activated by an Alignment tab the whole time.

A Number function on the Excel Ribbon provides data display choices such as controlling how often decimals are included along with numbers or any pattern of dates shown on the screen.

The Launcher button in the Number Category displays a Format Cells view box, which is activated by that of the Number tab.

In contrast to Font Group's manually formatting, the Styles Category proposes automated formatting options. MS-Excel conditional formatting commands are given by such solutions.

The Cells group contains commands for manipulating cells, columns, & rows, such as adding, setting widths, and heights& removing. MS-Excel also has options for hiding, shifting, as well as securing the entire Excel sheet.

Besides inserting & manipulating cell data, Editing has several popular commands. The AutoSum method allows you to quickly create simple formulas with only a few single or double taps. The Filter & Sort options make it simple to sort details vertically where the cursor lies. There are also options for deleting, finding, & replacing documents in this group.

Insert Tab

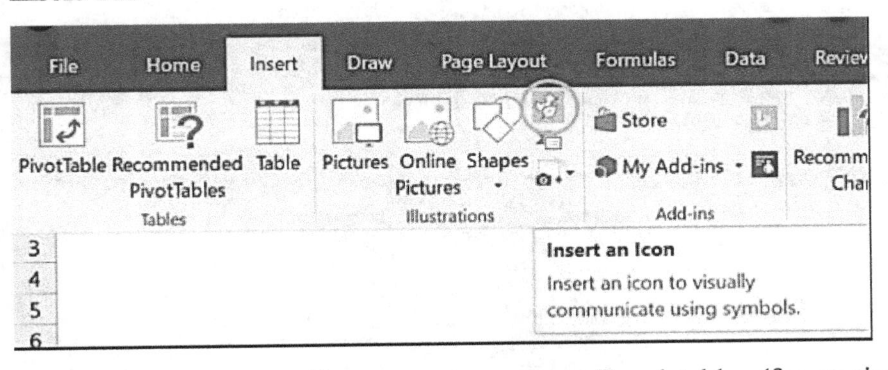

The Tables Category allows you to generate Excel tables (for storing column data) and pivot tables. Although the pivot tables store their data in the workbook cells, they prevent normal access to those same cells; those cells cannot be modified individually in the pivot table.

The Drawings Category within Excel allows you to incorporate digital images, including vector graphics. Photos can be dragged from even a hard drive or the internet, and Bing picture searches are also available. The figures & Smart Art keys in MS-Excel can be used to create vector images. The Screenshot function allows you to capture every detail of any window currently open on the monitor screen.

MS Office Solutions software can be found therein the Excel Ribbon Applications range. Office apps are the same as Android & Apple apps: they're just apps that Excel (or other Office software) can handle, and they have a useful function or usefulness that isn't built into the program.

The Excel Ribbon Charts Section enables you to link Excel & pivot diagrams to every workbook. A pivot map seems to be a chart that uses a pivot table as a database instead of using the regular Excel data cap.

The Reports Category of the MS- Ribbon contains only one item: the Power display symbol. The automatic Excel add-in which must be allowed is Power View. This enables you to generate stunning market visualizations using the most recent Excel

Data Model

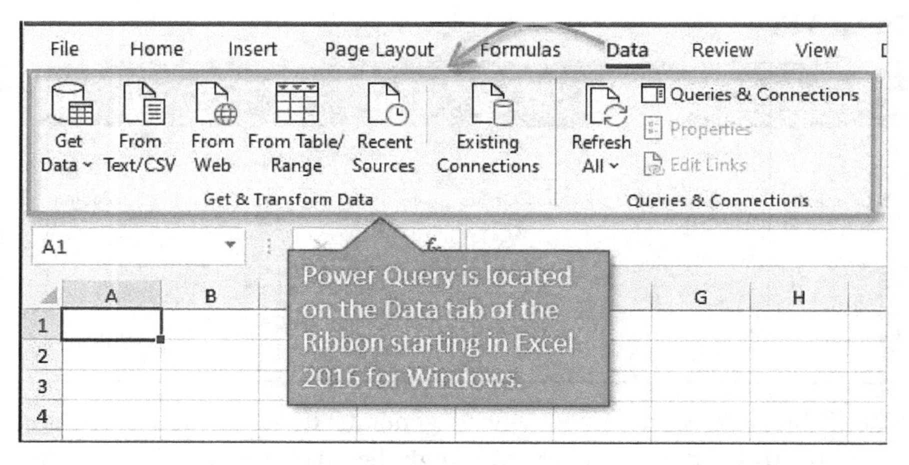

You can use the MS-Excel Ribbon & also the Sparklines label to add sparklines to the workbook. Sparklines are compact maps that appear in cells, usually next to data, and provide useful visual cues for patterns.

The Filters category within MS-Excel Ribbon contains buttons for connecting slicers & timelines to maps, columns & pivot tables. For filtering assembled data, those two elements have quite a highly engaging structure. Slicers can be used to filter any type of data, while timelines are used to filter dates.

Ribbon in Excel Only the Hyperlink click is included in the Ties category. Hyperlinks work in a similar way across all MS Office items.

You can add a hyperlink to almost any cell or group of cells in MS-Excel that will take the user to another workbook, open a different workbook, create a new mailing address, and open a website.

The text allows one to attach text-based objects that are located just above the worksheet's cells upon on an entity's layer.

There are options for inserting formulas & symbols in the Excel Symbols list. Symbols can be used in the same way as any other character in a cell. On the other hand, calculations are inserted within the text box near the top of its worksheet's cells on a piece's substrate. Excel has a detailed calculation toolbar with loads of arithmetic & science signs & forms to customize calculations.

Page-Layout Tab

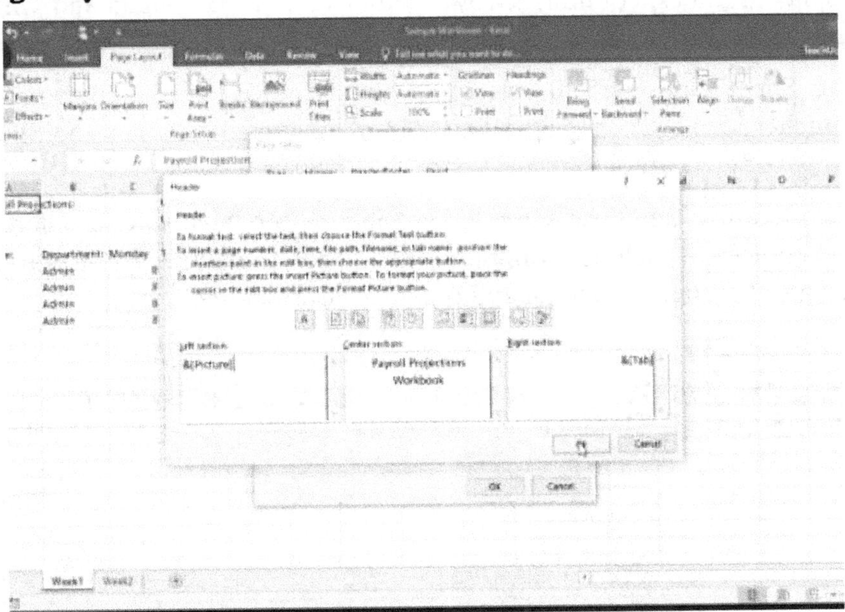

The Themes Group helps one to use the standard Themes feature found in all MS Office items. Themes provide a quick and easy way to monitor the overall layout of a post. When you choose items from the features, Colors, Icons, & Effects menus, menu items from the features are dropped. Using trendy and well-coordinated colors, a group may alter the appearance of cell types & graphic materials with a single hit.

The MS Ribbon Page Configuration Section contains commands that are commonly used while printing a file, such as adjusting borders, orientation, & page size and defining the print area.

The MS Ribbon & Page Setup group includes a launcher icon that opens the Page Setup dialogue box and allows you to use the Page Setting tab.

A Scale to Suit Section contains the commands found in the Page Configuration dialogue's Scaling section.

A launcher icon is included in the MS Ribbon in the Size to Match Section that displays its Page Config dialogue box, activated by a Page tab choice.

The Sheet Options Tab, as well as the Excel Ribbon, can hide and display gridlines, column, & row headings on all computers and while printing.

The Sheet Options Tab, including the Excel Ribbon, both have a launcher icon that opens the Page Setup dialogue box when the Sheet tab parameters are selected.

The Organize Category & the Excel Ribbon both have several different Tab choices. It includes commands like Sends to Back, Carry to Front, or Party that help you align and organize your objects.

Formulas Tab

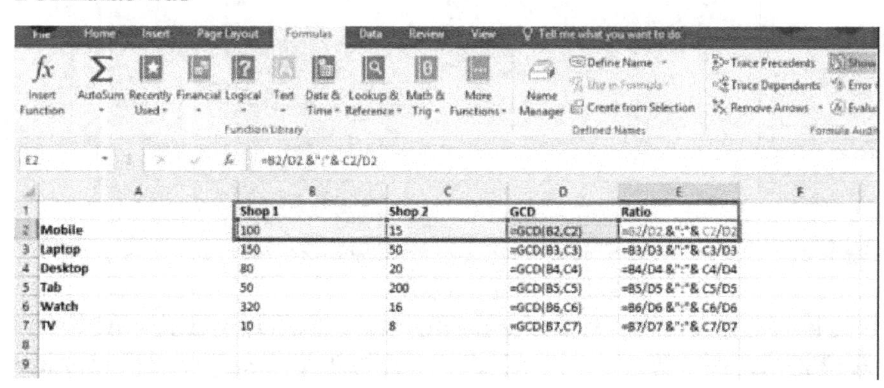

The formula Category option allows one to connect most of Excel's several hundred formula tasks. The functions are divided into classes, with its final stage (Further Functions) disclosing all categories without a decline-down menu in a submenus series.

There is no need for a launcher logo in this category since the main formula-linked dialogue (Insert Structure) is accessible via the final drop-down option in each of that drop-down choices.

The Identified Names group on the Excel Ribbon is related to the creation and management of names. On the other hand, Names is merely an Excel tool that allows you to link a text (name) mark to a collection of cells, an equation, and perhaps a static (taxation rates) attribute. You should use the name of the variable rather than just the value it represents when creating a formula. This ensures that the equations are consistent, and it gives you a simple way to change the meaning of many formulas by changing the meaning of a single phrase.

The Excel Ribbon, including the Formula Auditing section choices, are designed to assist you in identifying formula flaws. You can quickly highlight the cells to which the formula corresponds, and similarly, or step through a complex formula sentence through sentence using these options.

MS Ribbon, The Computation group, contains commands that track how and when Excel calculates. In worksheets containing multiple equations formulas that link to external workbooks, it is necessary to calculate formula results that may have a negative impact on the functionality of the worksheet. This category then offers the option to make a calculation manual, limiting Excel's ability to perform calculations only.

Data Tab

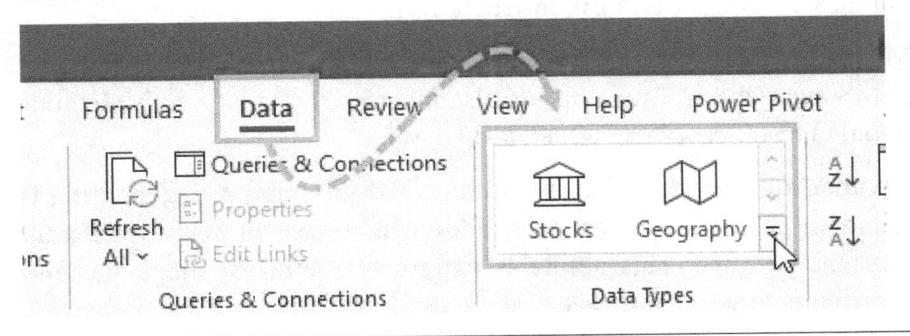

A Get outside Data panel contains commands for importing data into Data Structure from a variety of sources, which can then be utilized to perform data analysis & visualization through pivot formulas, the Power View, and pivot charts choice. These features are also available in PowerPivot choice, which provides a wide range of possible functionality for modeling collected data sets.

MS Ribbon the Links option has ways for dealing with existing connections A Worksheets Connections dialogue appears when you click the Relations icon, and you can check for & uninstall links. The Refresh All option allows you to replace old editions of downloaded files, which are stored forever inside a data structure, by the most recent edition from the author.

Microsoft Excel Ribbon You can use your Sorts & Filter category commands to perform basic and sophisticated sorting as well as tabular data processing. The Sorts from A to Z or Sorts from Z to A buttons facilitate one-click sorting across a single column, while the Sort icon displays the Sort menu, which allows for multi-column, color, and customizable sorting.

Similarly, the AutoFilter key makes it simple to filter information based on predefined criteria, whereas the Specialized Filters function allows one to distinguish data that mostly meets somewhat challenging criteria.

The Computer Tools category gives one access to a number of advanced, unified MS-Excel resources, one of which performs a meaningful data manipulation task. E.g., the function Texts to Columns allows one to filter data currently in a single column into multiple columns (usually after it has been assembled in MS-Excel). There are also good Delete Duplicate Files & Flash Fill instructions in this group.

Microsoft Excel Ribbon Data analysis techniques that have become part of MS-age-old Excel's arsenal are also included in the Research Methods (Goal Quest, Scenarios, as well as Datasets).

Another long-standing Excel feature called grouping & highlighting is found in the Layout Category: adding a degree of self to different columns & rows. This ability is extremely useful for managing and controlling large worksheets.

Review Tab

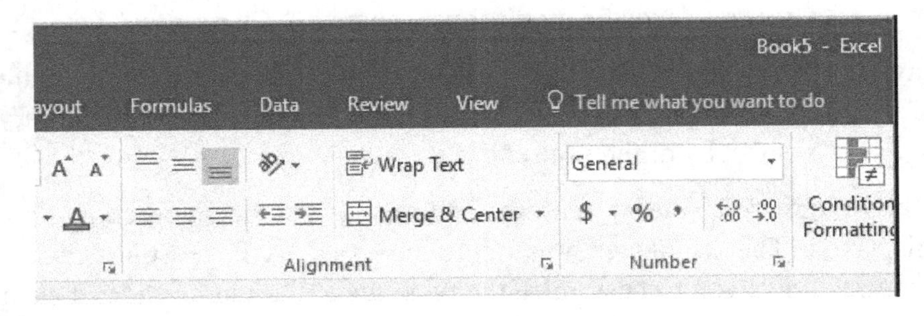

The Proofing tab involves the common Spell-checker, including Text commands, as well as a testing tool for conducting searches in a number of local and online tools such as Bing & Encarta.

Excel Ribbon The Language button is indeed a subclass within an analysis tool that allows you to translate common words and phrases into over thirty various languages.

Excel Ribbon The Comments feature lets you annotate responses on workbooks for your benefit so as a workspace feature, as well as share your thoughts on workbook content with colleagues. There are buttons for creating, uploading, deleting, publishing, or even hiding messages in this category.

Excel Ribbon Upon the Analysis Page, the Adjustments button has the most powerful commands.

The View Tab

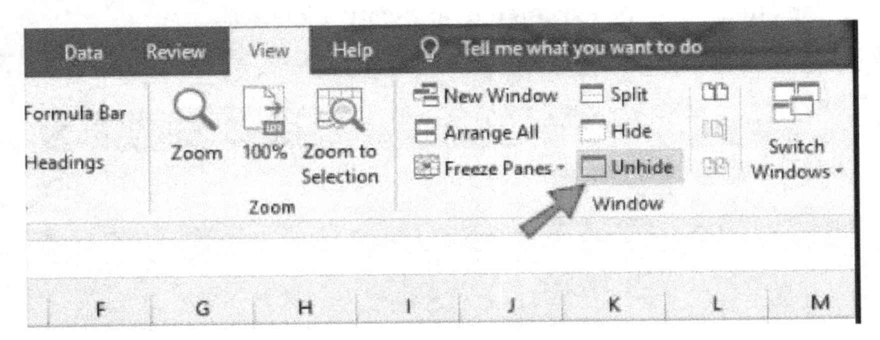

The Protective Sheet & Protective Worksheet controls may also be used to restrict access to specific monitor user changes, cell ranges & keep the layout of the worksheet intact. Several commands allow you to share worksheets with several other users and track changes applied to the very same worksheet by different users.

The Workbook Views feature controls how the workbooks are displayed: The regular view is made for speed. A Page Split Preview feature allows you to customize the page. Page Design provides a clear picture of how papers get printed.

Microsoft Ribbon The Display button contains commands for viewing and hiding the key elements of the MS-Excel GUI, such as column or row headers, grid lines, & formulation bars.

The Ribbon in Microsoft Excel The Expand tab keeps track of the magnification in work. For instance, you might highlight a range of cells and use the Maximize for Selecting key to maximize MS- Excel such that the selected range fills the screen.

Excel Ribbon That Windows Type offers solutions for dealing with a wide range of documents. For instance, the Organize All command will ease and scale all currently open windows so that you can just see their contents simultaneously.

Excel Ribbon The Modifiers feature also has a popular drop-down command that allows you to see most modifiers in all open worksheets and build custom macros. (A macro is simply a series of automated instructions that could be executed by pressing a key.)

Help Tab

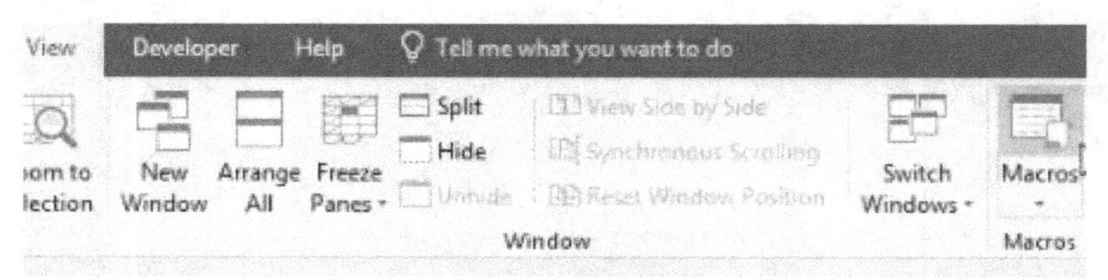

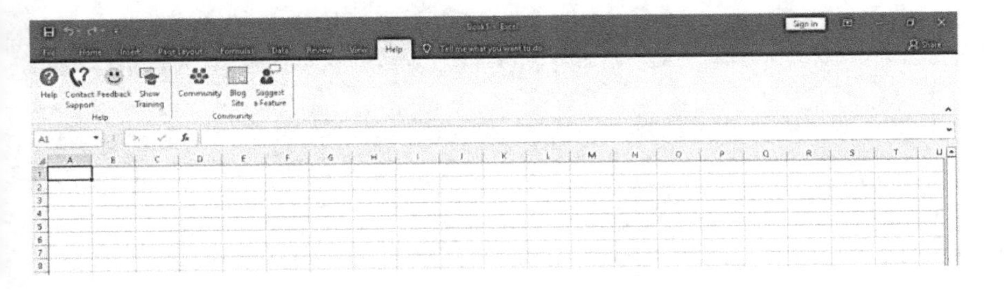

Across the whole of 2018, a new Support icon appeared in MS-Excel. In MS-Excel 2k7-2k16, the MS-Excel 2k3 Support menu remained limited to the circled question icon in the upper right portion; Support is now a full-fledged ribbon tab.

First-ever key is MS-Excel Online Support. MS Excel consumers have always resented the switch to online support & have fond memories of the good old times of offline help. However, online support has been improving over time. Each month, over 50 million people have viewed over 2,000 web aid posts.

Feedback allows one to offer up a screenshot along with a message on what you prefer & don't like. This suggestion is continuously being read by the Excel team.

The view guide opens a support panel featuring a selection of recent videos that have been integrated into Excel.

Chapter 7: Using Microsoft Word To Generate A Spreadsheet

Though MS-Excel has been the MS Office system and is best known being a spreadsheet program containing a lot of features, there will be times when one needs a spreadsheet within a company report, including a bunch of Txt files. By adding it as an object in the text, you can create an Excel spreadsheet in any other Word document. Word makes the process easier by allowing you to generate a spreadsheet in a variety of ways. You can choose from the Insert Item menu or the Table menu.

7.1 Recognize the Microsoft Word user interface

MS-Word provides customers the ability to merge excel data effortlessly into a single word document. The Insert tab enables you to easily add excel info. MS-Word is indeed a word processor that allows one to create a variety of pages as well as other types of data, including emails, flyers, & studies. With the addition of several enhancement features, such as the ability to create and work on online papers, Word 2021 gives you the ability to do anything, including any word processing task.

When one opens Word 2013-2021 for about the first time, A Word Starting Screen appears. You can now create a new document, choose a template, or open an existing one.

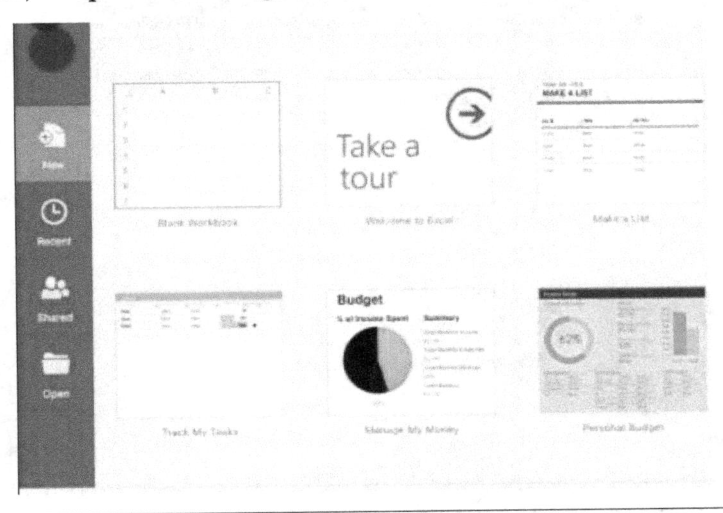

Word 2k13, 2k16, and also even 2k21 could sound familiar if you've encountered Word 2k10 or 2k7. The Ribbon & the Quick Access main menu are often used, and commands to perform simple Word tasks—such as that of the Backstage display—are available.

Rather than just a traditional menu, the word in 2k13 uses a tabbed ribbon structure. There are several tabs on the Ribbon, like one with a selection of command groups. These menus can be used to perform the most popular Words functions.

And, for the moment being, we only have the insert tab, so excel can only be used for this tab.

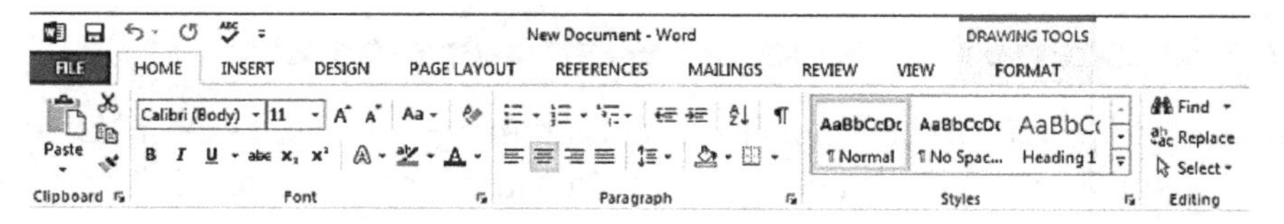

Choose the Insert tab by clicking or pressing it.

Begin dragging the pointer to the location where you want the table to appear in your text.

Choose the table symbol.

A pull-down button is available

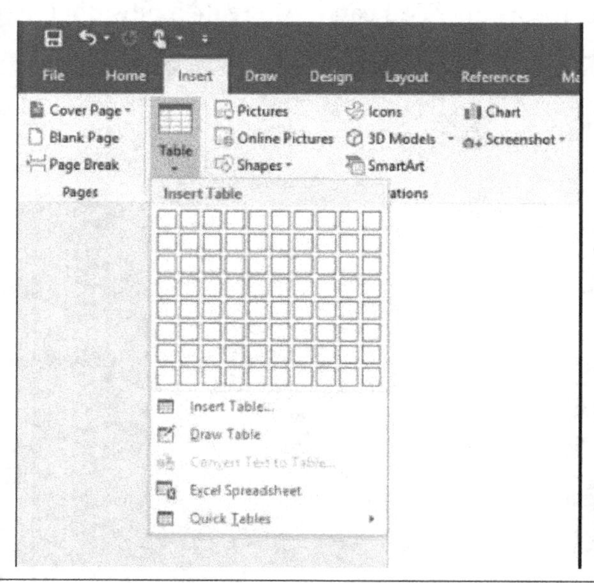

After that, end up choosing the Excel spreadsheet. Also, pay attention to the directions given below.

7.2 Import a spreadsheet into MS Word

To insert Excel into a word document, follow these steps:

Tap the Insert button.

Pick the Item command from the Text Category.

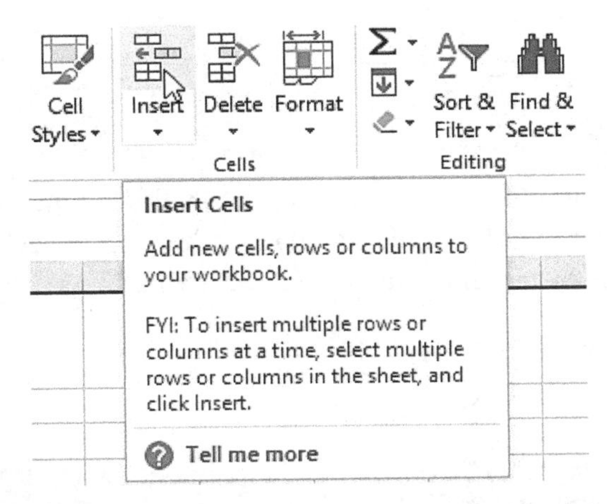

A dialogue box will appear. Select Build from its File tab, then Search.

Find & select the Excel file one would want to use, then hit Insert.

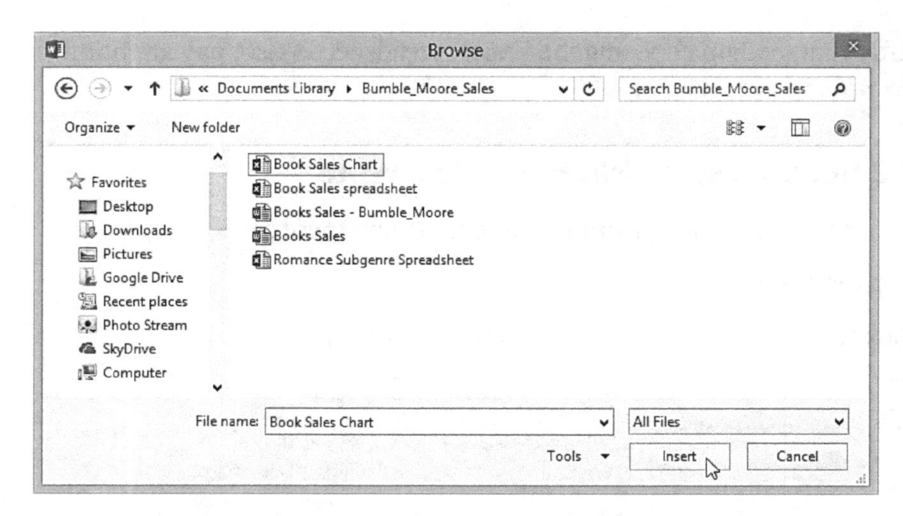

If you'd like to add the information to its MS-Excel table, check the box next to the document's route. As changes here to the MS Excel chart were made, the Word map will automatically update.

Go to the OK tab.

The Wordbook will screen the map.

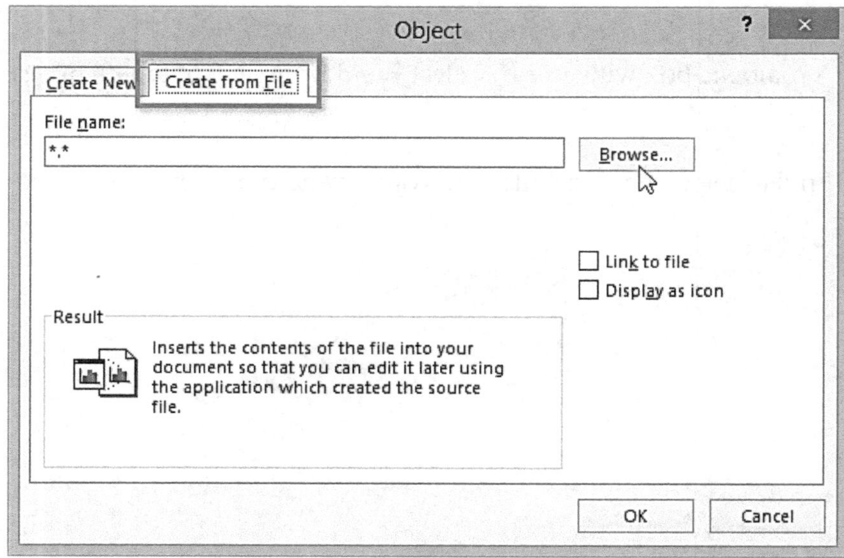

Further for editing a built-in chart, double click that map. The source data for that map will appear inside an Excel spreadsheet. Please save the Excel map once you've finished editing.

Chapter: 8 Excel Chart, Their Forms, And How to Create Them?

We need Excel to store info for small & large businesses as well as personal data. Even though spreadsheets seem to be necessary for data processing, these are cumbersome and do not provide a clear view of data trends & relationships for team members. MS-Excel can assist them in converting spreadsheet details into graphs to create an intuitive research report as well as make rational business decisions.

MS-Excel 2021 helps you to build graphs & charts for almost any purpose. If you've created an MS Excel chart or graph, you could use the Template button to modify and adapt it to your specifications. Learn how to design a map in Excel 2021.

8.1 Here are the distinct forms & types of charts?

An Excel graph is a graphical representation of details in bars & several other shapes. It's a visual representation of data from such a workbook which could help you understand the data better than just gazing at the figures. A chart seems to be a powerful tool for visualizing data in a variety of formats, including Bar, Pie, Column, Doughnut shape, Graph, Zone sort, Radar graphs, Scatter dots, & Floor graphs. Using Excel to design a chart is a fast and effective method.

8.2 The pie charts

In certain cases, pie charts have been used to show the individual significance of different values while still adding the total value. A single collection of data is often used in a Pie chart.

To build a pie chart on some kind of data set using the 2019 or 2021 versions of Excel, simply follow the instructions.

Pick the data set first from the A1:D2 range.

	A	B	C	D	E
1		Bears	Dolphins	Whales	
2	2017	8	150	80	
3	2018	54	77	54	
4	2019	93	32	100	
5	2020	116	11	76	
6	2021	137	6	93	
7	2022	184	1	72	
8					

Over the Design tab of the Chart's type, pick the Pie logo.

Click Pie from the menu.

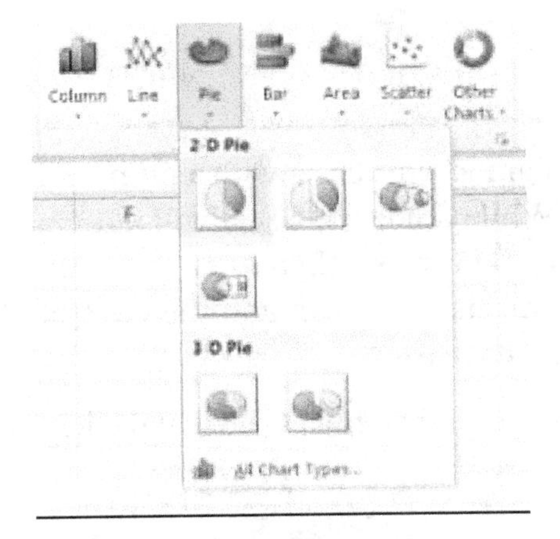

As a consequence, you will note the following.

Press on the pie so as to choose the whole pie. To take a section of the map away from the midline, tap on it.

Results:

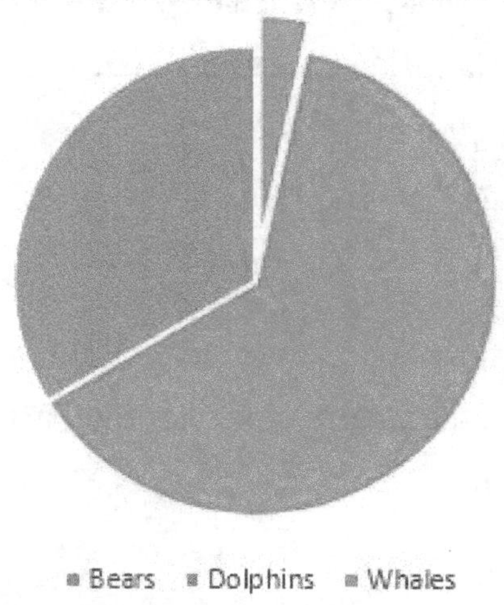

Note: If you're designing a pie chart with a numeric mark, make sure cell A1 is blank first. While doing so, Excel will not treat the figures within column Too as a dataset &, therefore, will build the correct table automatically. If you like, you can add the content Year during step 1 after you've made the Chart.

Pick the A1:D1 category, then hold CTRL and click the A3:D3 category.

	A	B	C	D	E
1		Bears	Dolphins	Whales	
2	2017	8	150	80	
3	2018	54	77	54	
4	2019	93	32	100	
5	2020	116	11	76	
6	2021	137	6	93	
7	2022	184	1	72	

Press Delete on the icon at the bottom of the screen (following the previous step of adding the graph).

Pick a good pie chart.

Pick a Data labeled check box by clicking the + icon on the desired section of the Chart.

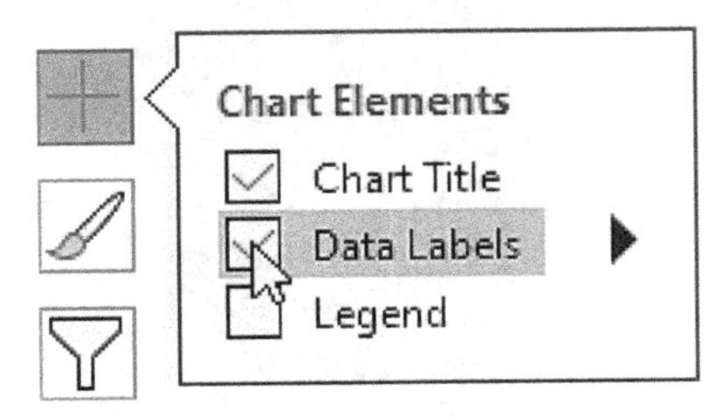

On the appropriate side of the Chart, hit the paintbrush icon to change the color theme of the pie chart.

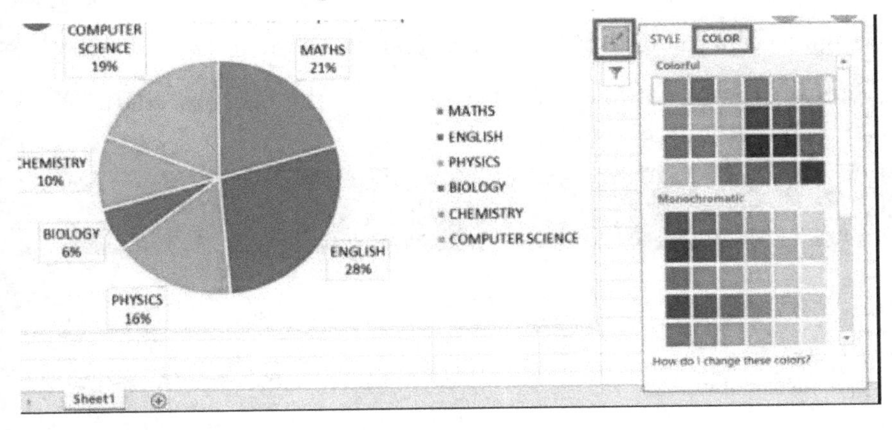

Right-click, that bar graph, then select Data Label Type from the context menu.

Uncheck value, verify Percentage, then press Middle to evaluate Name of Type.

As a result, you'll get something like this.

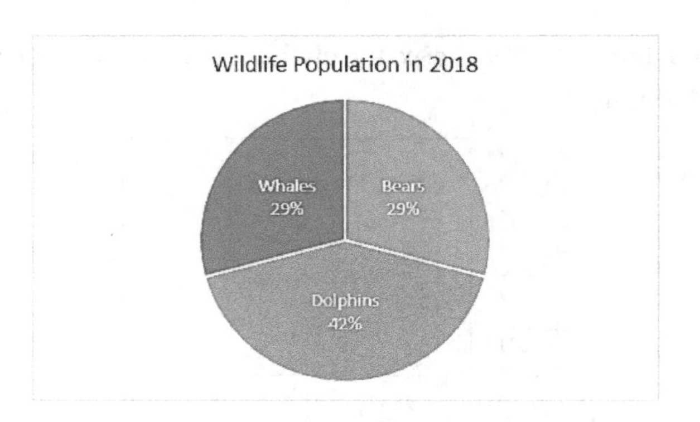

8.3 Column Graphs

To design a column chart within Excel, you'll take the following steps:

Highlight the information you wish to use in the column chart. We picked the range A1:C7 for this case.

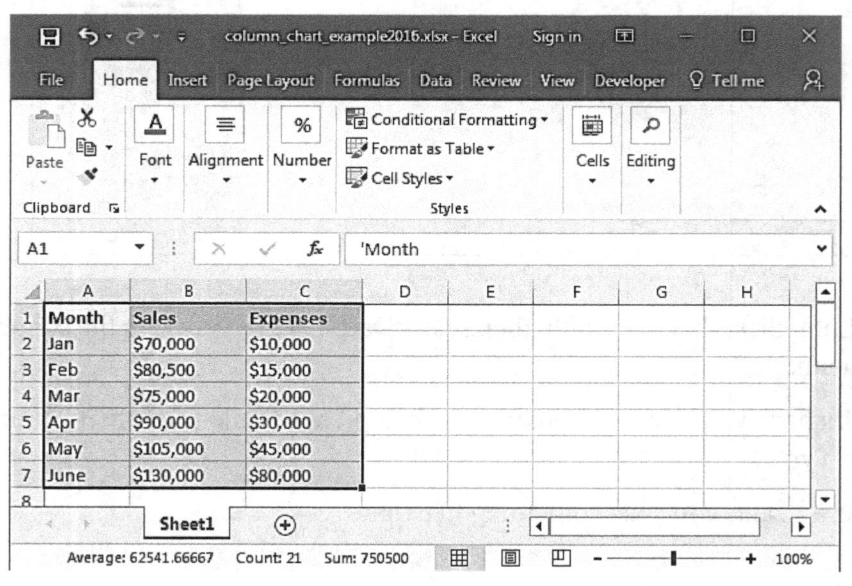

There in the toolbar at just the top of each page, select the Insert button. Click a chart from its drop-down menu by pressing the Excel Column chart option within Charts Category. For this case, we went with the very first column chart there in the 2-D Column section (regarded as Clustered Column).

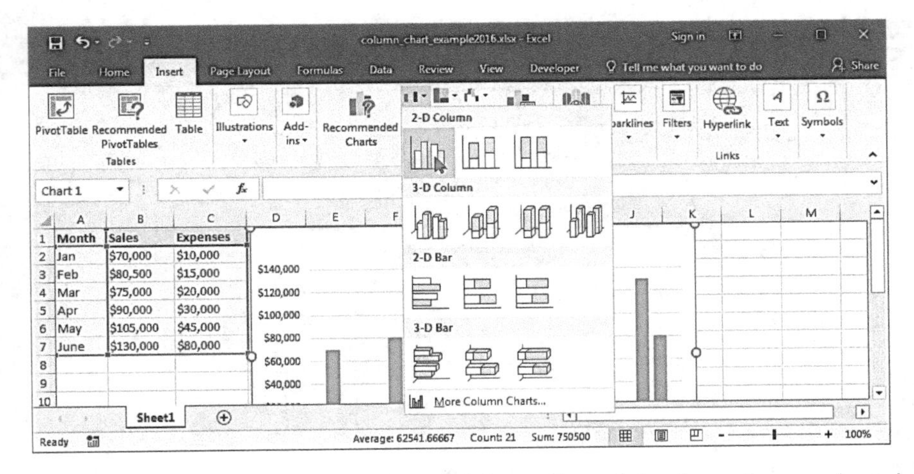

All sales and expense statistics will be reflected in the column chart in the rectangle bar spreadsheet. Vertical blue lines represent sales values, while vertical orange lines represent costs. For certain vertical lines, their values of axes could be seen along the left side of the graph.

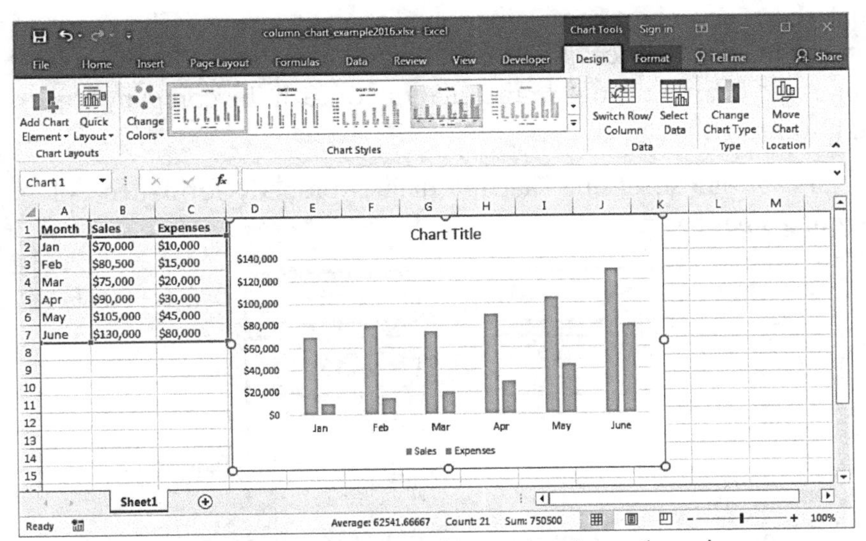

Last but not least, let's alter the column chart's explanation.

Click the "Chart Title" link towards the top of a graphic item to adjust the title. One ought to be able to notice that the title can be changed. Put the text you want to appear as the title here. Throughout this example, we will use the term "Sales & Expenses" to build a column chart.

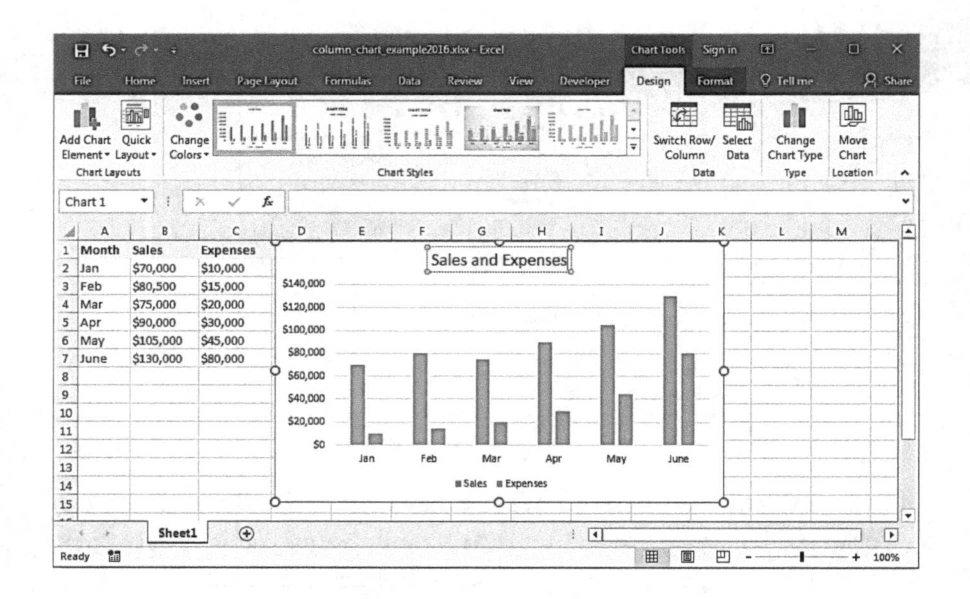

8.4 The Line Charts

A bar chart, or a chart containing long bars, is a form of Chart used to visualize the significance of data over time. For example, the accounting team might chart a change in the quantity of cash the company has on hand over time.

First and foremost, ensure that the data is properly formatted before designing a bar chart.

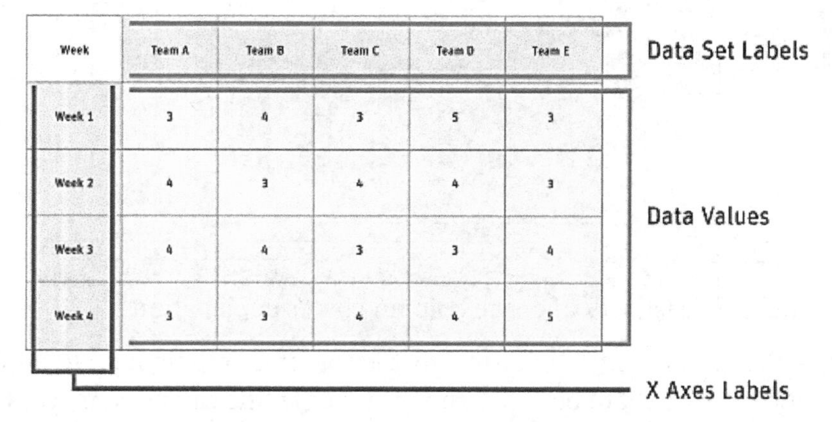

Week	Team A	Team B	Team C	Team D	Team E
Week 1	3	4	3	5	3
Week 2	4	3	4	4	3
Week 3	4	4	3	3	4
Week 4	3	3	4	4	5

Making use of Smart Draw Add column, pick line graph from the Graph menu.

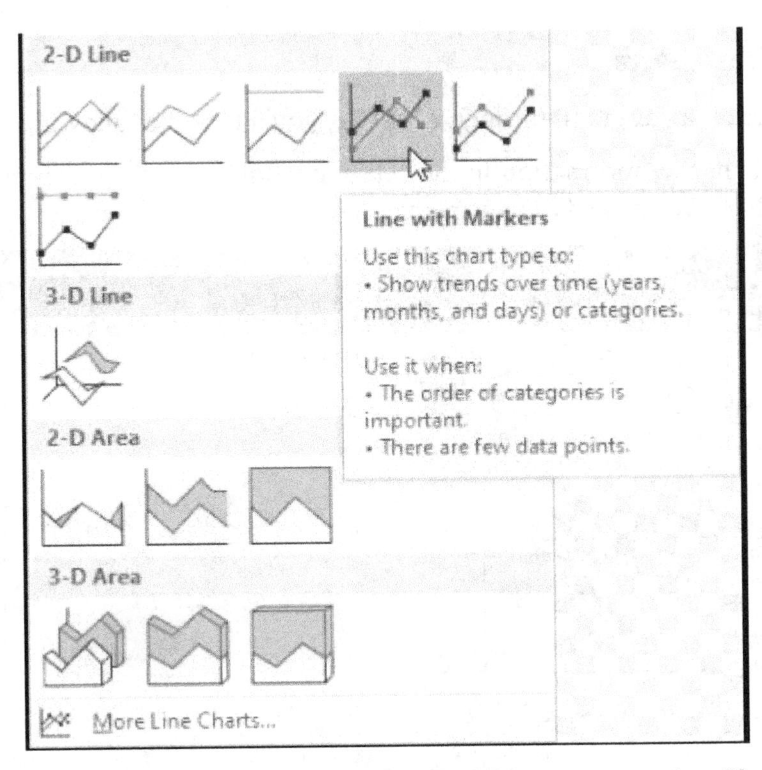

Select the information file you have to use to generate the Chart, & Smart-Draw will create it for you instantly.

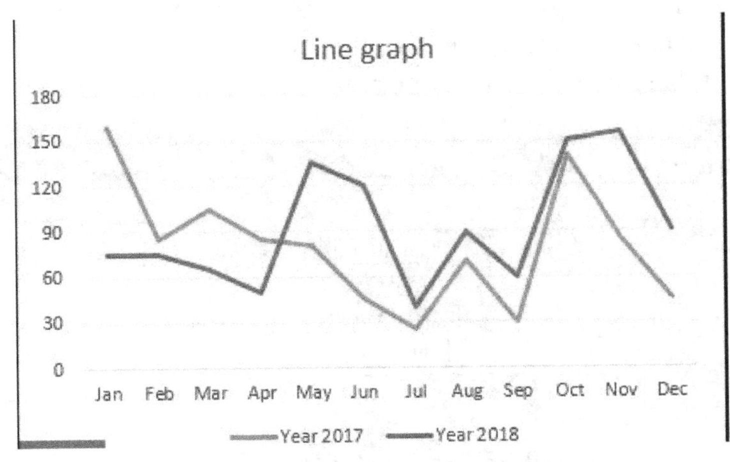

Utilizing Edit Graph tools & double-clicking an imported graph, one could easily alter the legend placement, description and modify the shape of a graph.

8.5 Bar charts

To design a bar graph through Ms 2021, you'll take the following steps.

Highlight the information in the cell one would need to use to make a bar chart. We picked a range between A1 to C5 in this case.

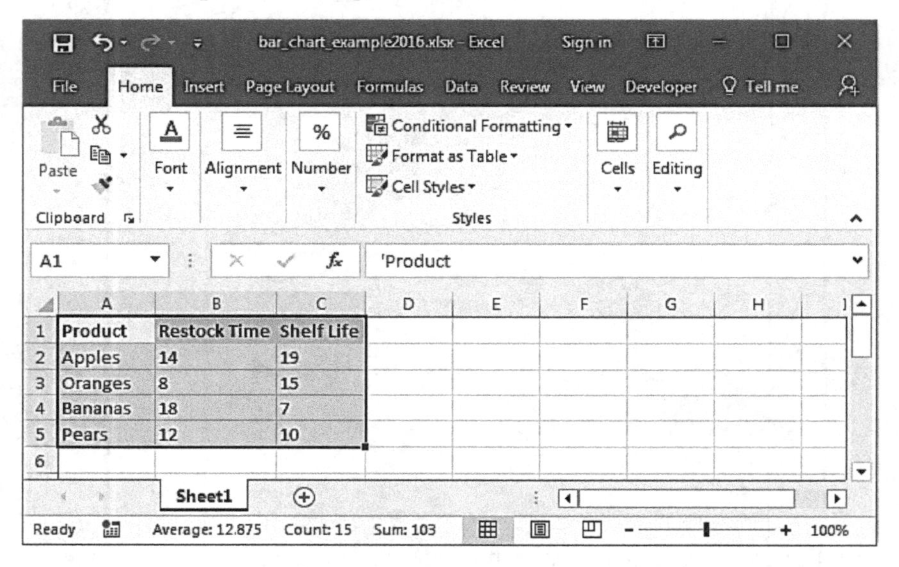

Within the toolbar, just at the top of every page, click the Insert button. Select a chart through the drop-down menu by pressing the Excel Bar Chart icon in the Charts group. For this case, we went with the very first bar chart there in the 2-D Column section (regarded as Clustered Bar).

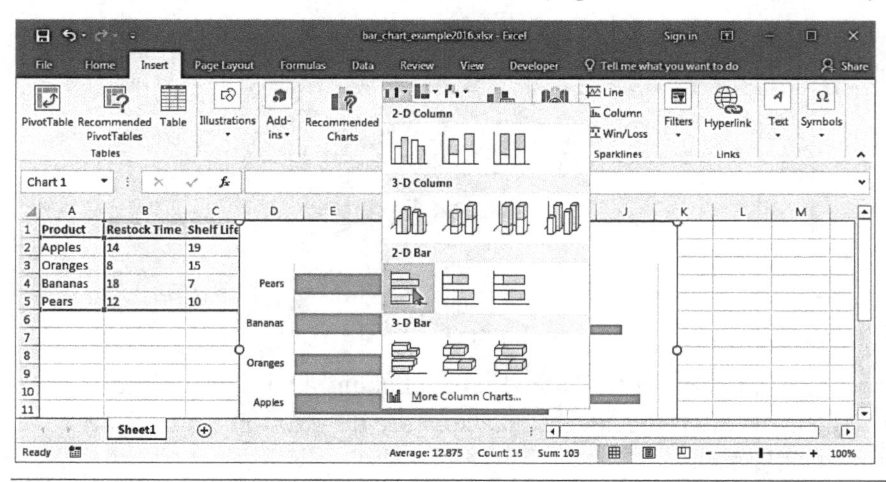

You can also use the bar chart as in horizontal bar spreadsheet, which shows the retail as well as the shelf life of each product. Shelf life is represented by orange horizontal lines, while retail life is represented by blue horizontal lines. Through such horizontal lines, one could notice the axes' values at the bottom of the display.

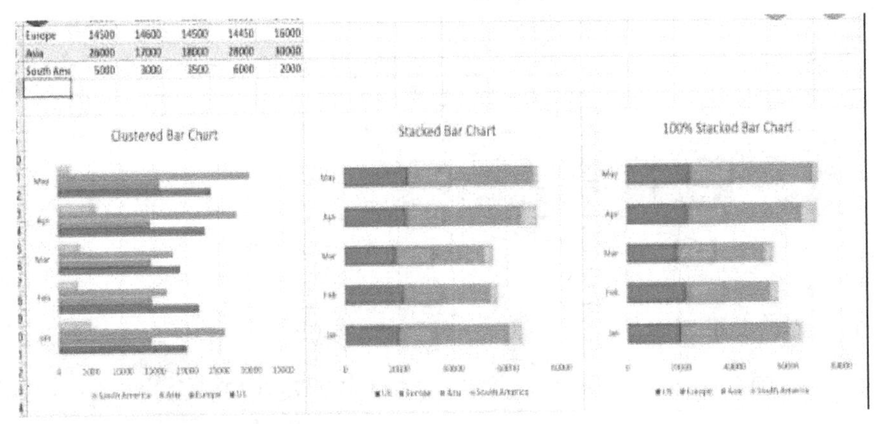

Eventually, let's update the title of a bar graph.

Hit the "Chart Title" icon near the top of a graphic item to change the title. You ought to be able to recognize that the title can be changed. Put the text you want to appear as the title here. Throughout this example, we'll use the bar graph term as "Product Life."

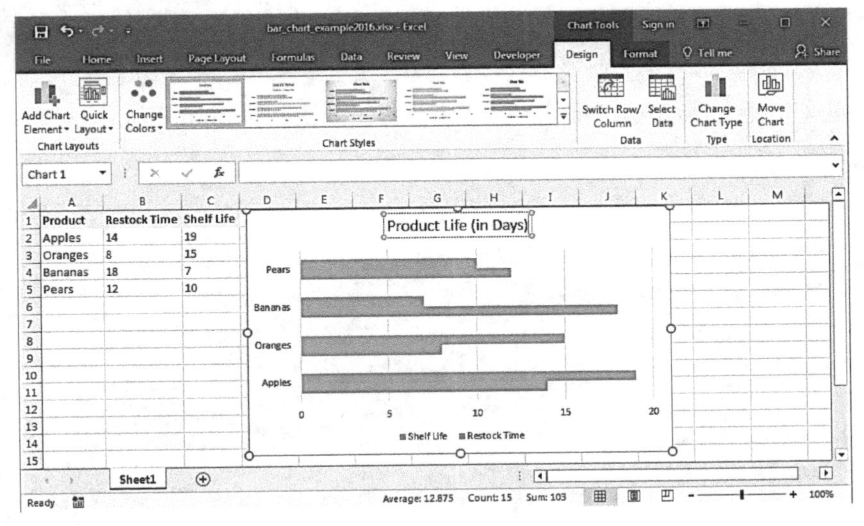

8.6 The area charts

The same guidelines apply to Excel graphs & charts. Let's look at an example and see how to design an Area chart.

For the area, we possess smart quarterly revenue results.

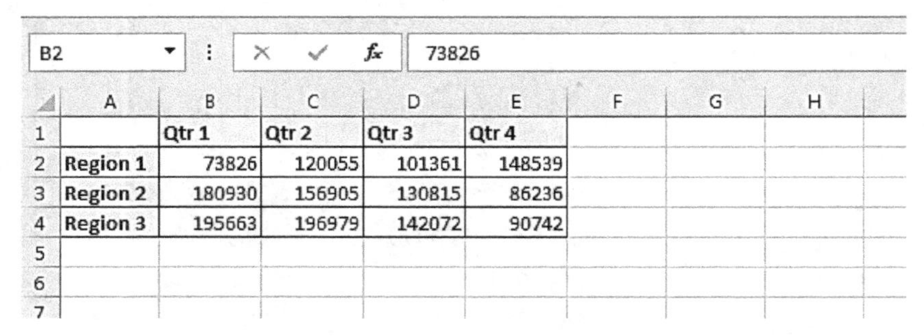

B2 | 73826

	Qtr 1	Qtr 2	Qtr 3	Qtr 4
Region 1	73826	120055	101361	148539
Region 2	180930	156905	130815	86236
Region 3	195663	196979	142072	90742

Choose the details

Go over to the Design category > then to Charts group > Select the Field graph.

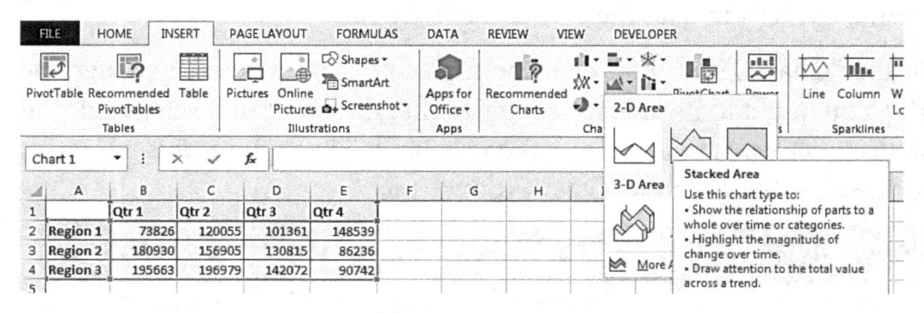

Pick Clustered Area Graph through those in the region graph.

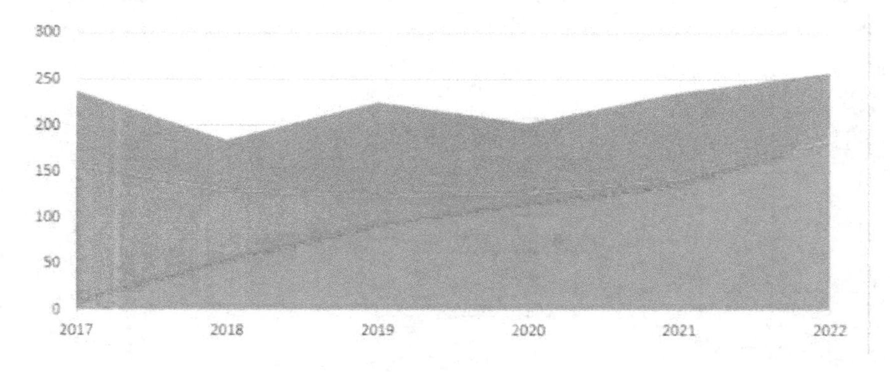

8.7 Scatter charts

Choose a selection of worksheets from A1 to B11.

Upon this Insert page, hit the XY chart (Scatter) order logo.

Choose a graph subtype that would not include a graph.

The data within the XY table (Scatter) can be shown in Excel.

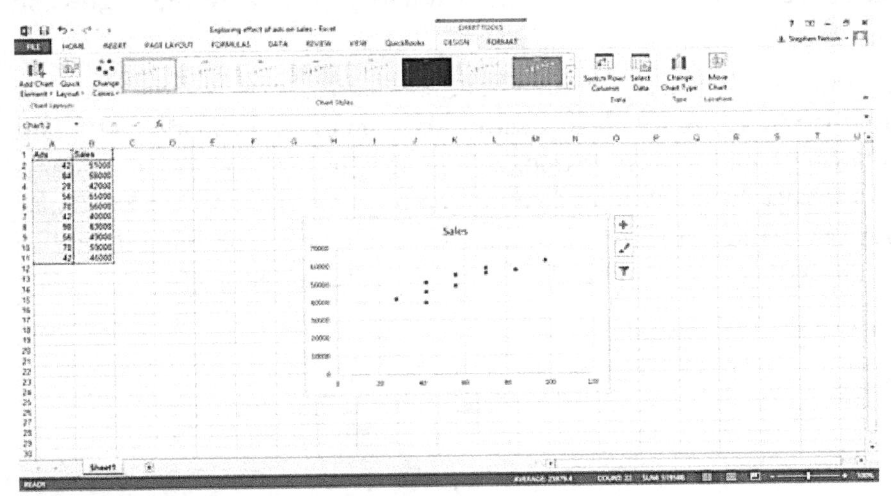

Double-check the Chart's data organization.

Check if MS has properly arranged the data by looking at the table.

On a Chart Tools Interface link, press the Turn Column/Row control key if you're unhappy with that of the graph's data entity- the data is backward or flip-flopped. (You can also experiment with the Turn Column/Row feature if you think it would be useful.) It's worth noting that the data is well organized. Increased advertising tends to be linked to increased sales, as seen in the graph.

Annotate the Chart as necessary. Make the Chart more attractive & readable by attaching those tiny blossoms to it. E.g., you might create a chart with a title or a description including the Chart's axes by using the Chart Title & Axes Titles controls. You can add a trend line by tapping the Add Chart Options menu upon a Trendline command icon. Pick the Interface click & then Add Chart Element option to display the menu of Add Chart Element. To access the Template page, one should first select an inserted map item or view a graph sheet.

The Trendline menu would appear in MS-Excel. By pressing on one of several of the accessible trendline options, one can specify the number of trendline and correlation estimates one may need. For e.g., to run any linear regression model, click the Linear button on the keyboard. A Trendline Graph Tools Configuration Tab in Excel 2k7 is where you introduce a trend line. To the scattering plot, add the regression equation.

To monitor how well trendline multiple regressions get measured, just use the ctrl key & text boxes within the Trendline Format panel. Place Intercept = checkbox or textboxes, e.g., may be used to force a trend line to intersect the x-axis after a certain point, such as zero. One could also have Forecast Forward & Primitive texts to emphasize that a trend line can be extended beyond and even before existing data.

Tap on the OK button.

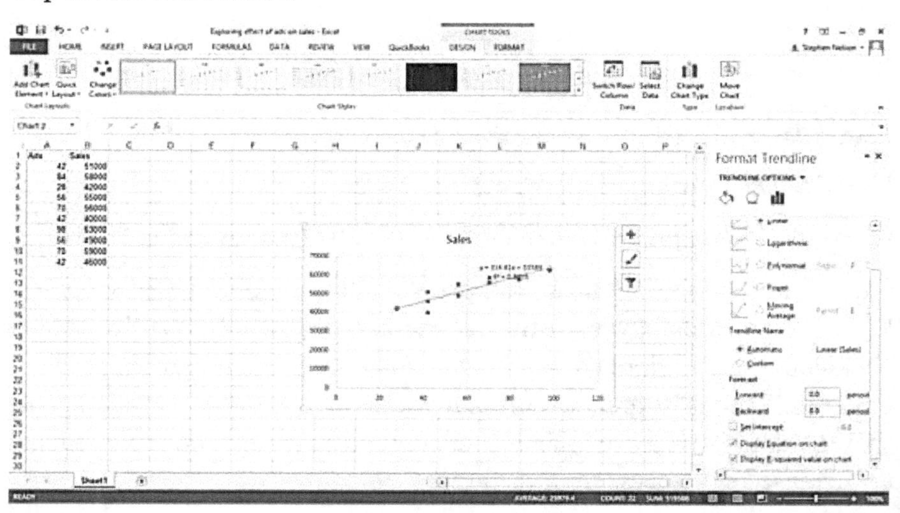

You could barely see the regression details, so it all has been annotated, making it a lot clear.

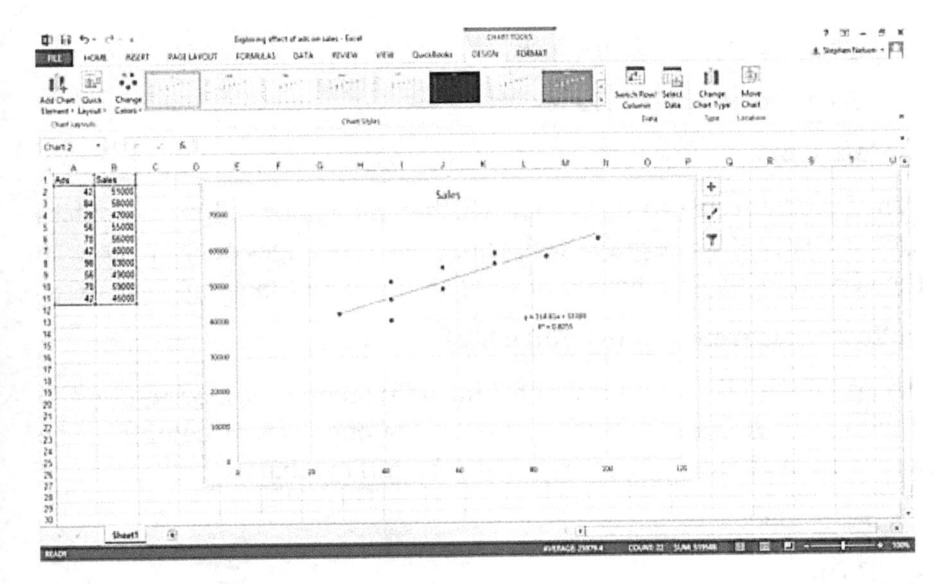

Other charts types

Bubble Chart

A Bubble graph appears similar to a Scatter graph, but with an additional third column to clarify the scale of the bubbles that depict data points therein data sequence.

The subtypes of the Bubble Chart are as follows:

Bubbles

A three-dimensional visual effect bubble

Stock Chart

Stock style charts, as the name implies, will show price changes in stocks. Nonetheless, the Stocks Chart may be used to show changes in other figures, such as average rainfall or annual temperatures.

Place data into rows or columns in a specific order onto a worksheet to make a Stock graph. To make a simple low high stock chart, for e.g., arrange the data with such as Low-High- Close insert like as Column Names in such an order. The subtypes of the Stocks Chart are as follows:

- High-low-proximity
- Amount of high-low-close

- Volume of Open-High-Close
- Open- closer-Higher-Lower

Surface Chart

When you're trying to figure out which combinations of two variables are the best, a Surface graph will help. Colors & shapes depict regions in the very same way as they do in what seems like a topographic chart.

Follow these measures to build a Surface chart:

Ascertain that almost all divisions, including data series, correspond to integer values. On a worksheet, sort data along rows or columns. The following subtypes are represented on the surface chart:

- 3-D surface area
- Contours
- 3-Dimensional wireframe layer
- Wireframe's contour

Radar Chart

The Radar chart compares the values of several different data series. To generate a radar chart, arrange information in rows or columns upon a worksheet.

The below subtypes are included in the Radar chart:

- Radar & Markers
- Radar Loaded
- Simple Radar
- Combo Chart

Such Combo graphs merge two or even more graph formats to generate data easier to understand, especially when there is a lot of it. It's visible from a secondary axis that's easier to read. To make a Combo table, put information in rows & columns upon a worksheet.

The variants of the Combo chart are as follows:

- Custom variations
- Panel Cluster – Line
- Secondary Axis Rows in a Grouped Panel
- Layered Field – Clustered Line

8.8 Excel Chart Customization

When the Chart Elements icon (with plus mark symbol) is selected in Excel 2k16-2k21, it presents a list including the main chart things that one can add to their Chart upon on the right-hand side of its built-in screen. To introduce an object to the table, hit the Chart Elements symbol to bring up a list of all Axis through Trendline in alphabetical order.

For e.g., to reconfigure the Chart's title, click on the Follow-up button on the Chart Elements toolbar associated to graph Title to display & choose between the below options upon its Follow-up menu:

- Include or reconfigure the chart title just above the plot area, centered above the line.
- Just use Focused Overlay Title to add or readjust the chart title at the top of a plot field.
- More tools for opening the Format chart. To adjust almost every aspect of title formatting, use the choices that appear when selecting the Rows & Fill, Effects, also Scale & Assets buttons below. Title Choices & Script Layout & Fill, then Text Effects, also Dialog box tabs below Text Options there in the task pane.

The research clustered column chart, including a data table linked, is depicted in the following image.

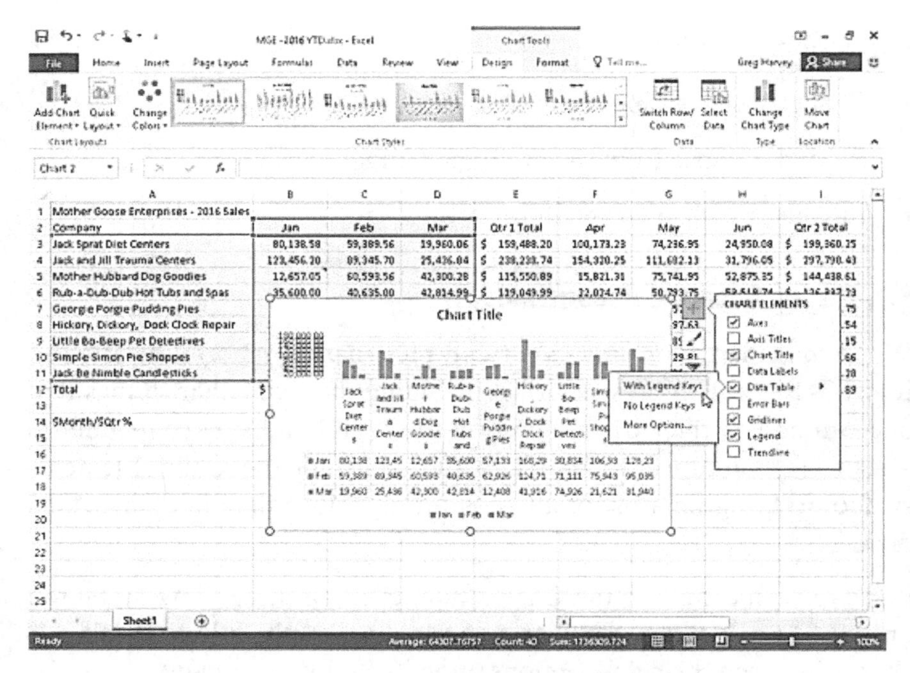

The first column of this informational table includes the legend's keys.

8.9 How significant are the charts?

MS-Excel has several automated tools, such as a graph function, to cope with all these other data storage values.

Anyone with accessibility to a spreadsheet could change data after it has been processed in the Excel database to view and express its significance. The chart function may be a key component of these systems.

Visualization

Spreadsheet managers may use Excel charts to create visual interpretations of data sets. Users can create various charts upon whom data is graphically depicted by highlighting a collection of data within the Excel spreadsheet & adding it to the graphing feature. Excel charts suitable for management/company presentations would help explain and convey the data set. A chart, rather than a table is containing lines of figures, which provide a clearer view about a set of data variables, allowing administrators to incorporate this interpretation into analysis or even plans.

Customization

MS-Excel simplifies the process of constructing charts from pre-existing data sets. Unless the spreadsheet has already changed data, the chart function might convert it to a graphic with just a minimum amount of user feedback. MS-Suggested Excel's Charts tool is an essential aspect of the phase. With only a few clicks, spreadsheet administrators can generate a chart, pick a chart type, and customize the names & axes.

Integration

When a business or other organization requires a database, data managed within MS Excel may be incorporated using the Excel chart function. For instance, whenever an Excel spreadsheet creates a chart using data in a worksheet, that Chart updates automatically as the data changes. This allows company managers & supervisors to keep track of their data as well as visualizations in a single feature, allowing them to quickly review reports.

Chapter 9: Formatting of the tables

9.1 Choosing a table style

Pick a table style to quickly format a large number of cells. You could also create your custom premade template, which will enable you to easily format any cell by specifying a cell type.

To do so, take the following steps:

Select all of the cells inside that data under the selection by clicking and dragging.

Pick "Format-as-table" from the Styles group upon its Home page.

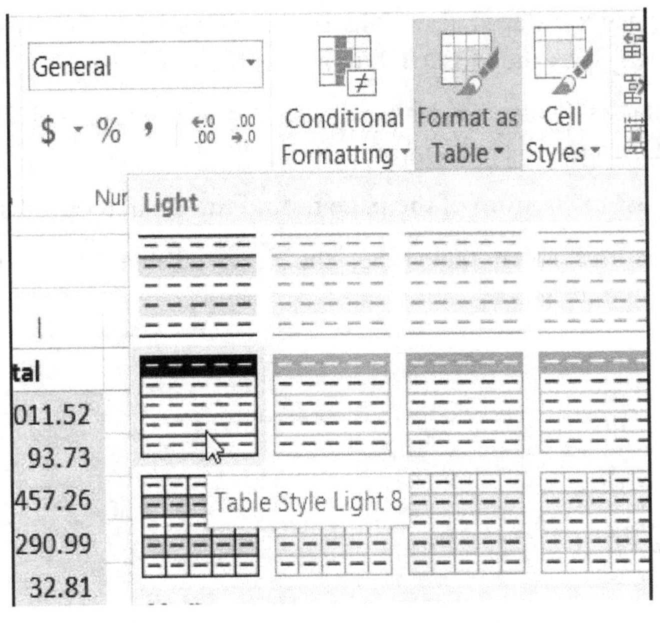

1. From there, one has the option of selecting whatever table type one would like.

Note: To make the first table design, push New Table Type or right-click the table design. To generate a new template that is similar to the original, tap Repeat. When you change the custom table design, it affects all of the tables throughout the worksheet using the style. This would help you save a lot of time.

MS-Excel would choose the data for one automatically. Please pick 'format-as-table', then click OK.

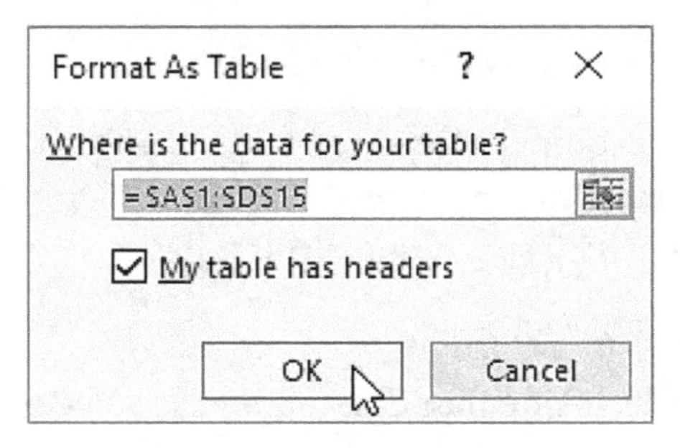

9.2 Deleting or designing a customized table style

Using the general steps below, MS-Excel will assist you in creating your table types:• Click the Home tab to access the ribbon.

Hit Format-as-Table tab within Styles Group. MS-Excel provides a variety of formats to choose from.

Scroll to just the bottom of the page to find the desired options.

Choose the most recent table type. The Newest Table Type dialogue box can be shown in Excel.

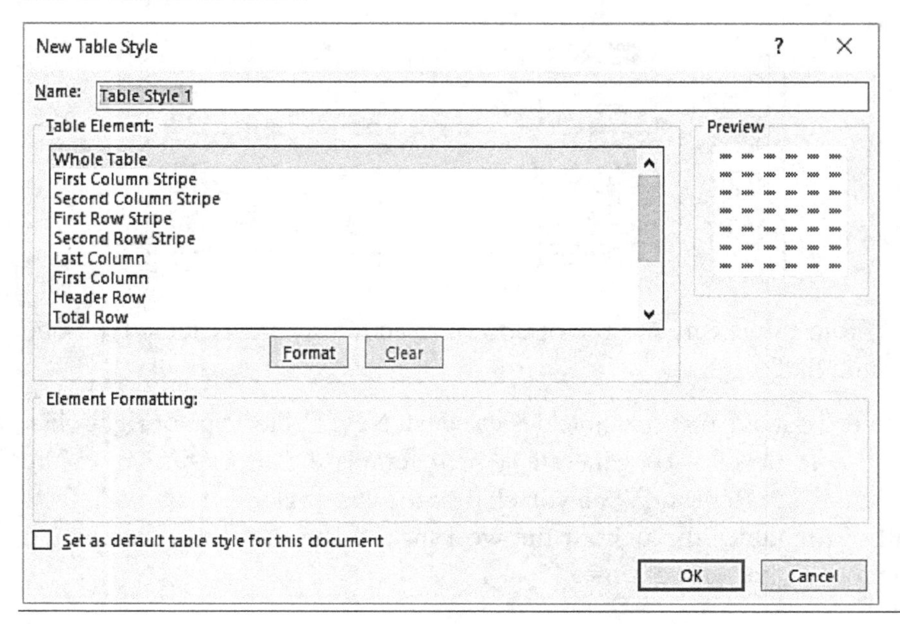

During this stage, give the customized table style a title.

Using the dialog controls to customize the theme's appearance.

Click OK to shut the dialogue box.

When you create a custom theme, it's automatically linked to that same Table Styles gallery:

To change the look of any table, just go to Table designs gallery, then right-tap on the theme, and pick "modify"...

Right-click on a custom table template, then pick Delete to remove it.

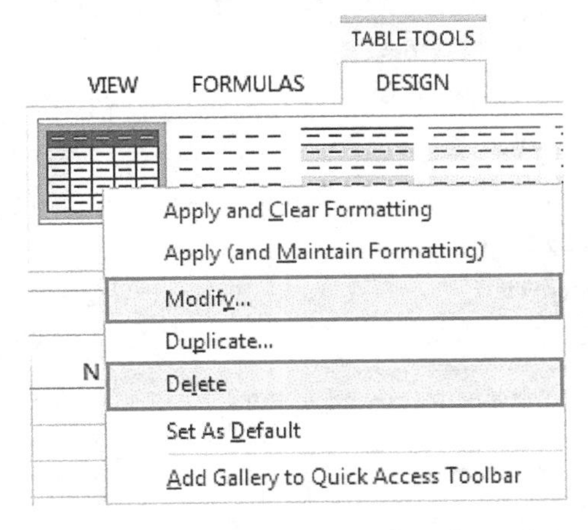

The Excel table's design component cannot be altered or removed.

Note: You can only use the customized table design in the workbook where it's been developed. The simplest way to add it to some other worksheet would be to copy a custom sort table to that certain worksheet. You may remove the copied table later, but the custom design would remain there in the Table designs gallery.

9.3 Specifying a table template to format the table's components

The primary benefit of formatting data here as a table is the variety of styles & filter options available. Click over a sign of "down-face-triangles" with header lines to locate them.

One may sort a table in three distinct ways from smallest to largest, then vice versa. This approach is completely self-explanatory. In a table, pick the choice over the top of a drop-down display by pressing a down-face triangle there in column one would wish to filter and sort.

One may choose a color to sort the results if one wants to utilize some background color to fill, use a color scheme manually, and even use conditional statements. We've taken the formatting choice to format its top 10% of query volumes using yellow fill, and thus it is bottom 10% using red font in the example below.

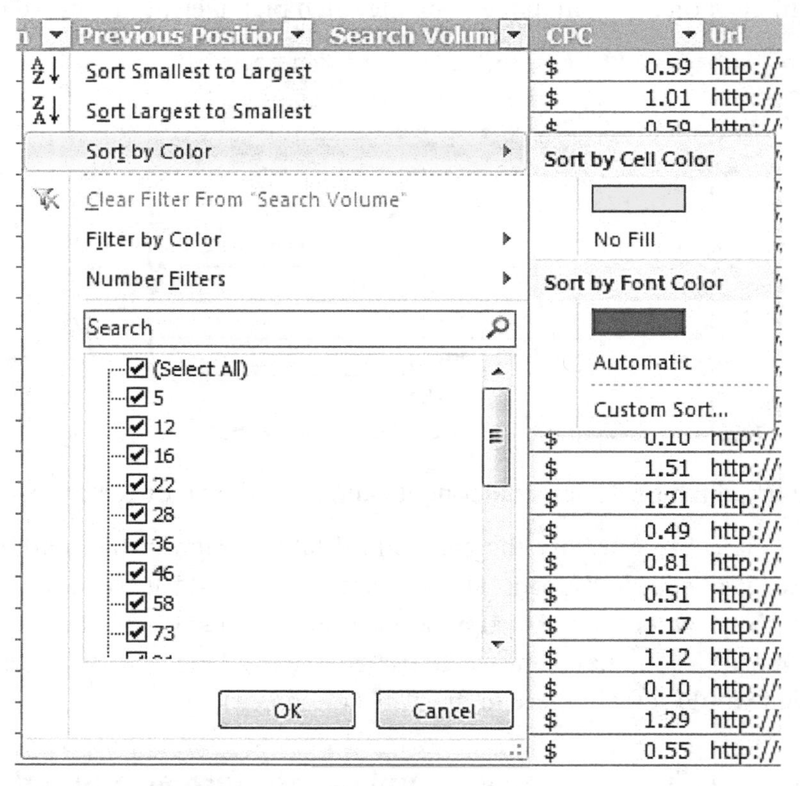

As a result, one can now use the Sort function to sort columns across cells by no fill, yellow fill, or red text, or blend the following. We would tap upon on yellow bar just below Cell Color Type if we wanted certain keywords with the highest Google Trends to float to just the peak of the table.

Options including Filter

The filter choice, since the name implies, will hide rows relying on the variables you choose. At this stage, no data is lost. It's just hidden for a few moments to help you fine-tune the data you're trying to decipher. Pick Transparent Filter here from its drop-down toolbar to the PC & Clear Filter icon on the Mac to release filters at quite a specific time.

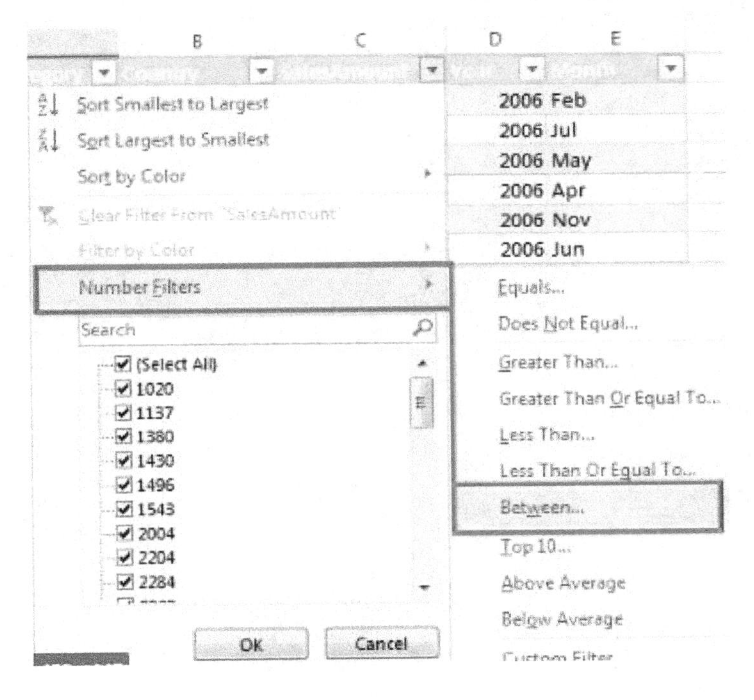

Chapter 10: Bringing a Signature to an Excel Spreadsheet

When a signature is added to an Excel file, no changes can be made to the file despite invalidating the signatures. This lets you and several other advanced customers aware that the file has been mostly viewed & accepted and that the author stands behind it.

10.1 Inserting signatures line & placing signatures

Launch the workbook in which you want to use a digital signature but keep in mind that none of the data will be altered or updated. Place the cursor near the point where you want to introduce the digital signature in the blank cell.

select Insert > then Signature Line > and also MS Signature Line Along the ribbon area. Take a look at the image below.

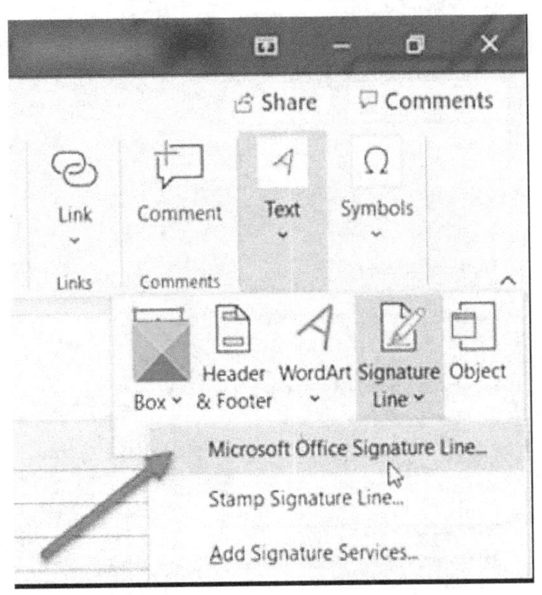

After tapping the Office Signature text in Excel 2k7-2k21, a disturbing background menu would appear. Simply select the option to "Do not display the notification again" and click OK to proceed.

Thus, some Signature Setup context would seem; follow the steps given from here on out:

For signature purposes, type the name in the designated text box.

In the assigned signer's name texting box, type the title;

Fill in the required signer's mailing address texting box with the email address;

If you wish to write a comment to the signature text, hit the Allow icon in the Signed dialogue box;

To exit the Signature Setup context toolbar, tap OK.

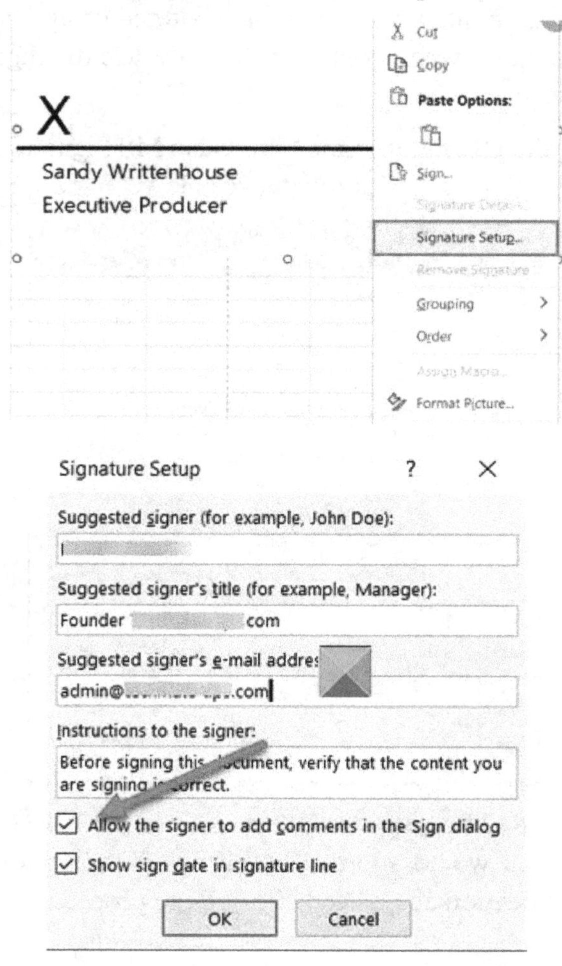

The signature text is then added to the proximity of a cell by MS-Excel through a graphic object. At step 2, put the pointer inside a larger X symbol that included your name & text. Take a look at the picture.

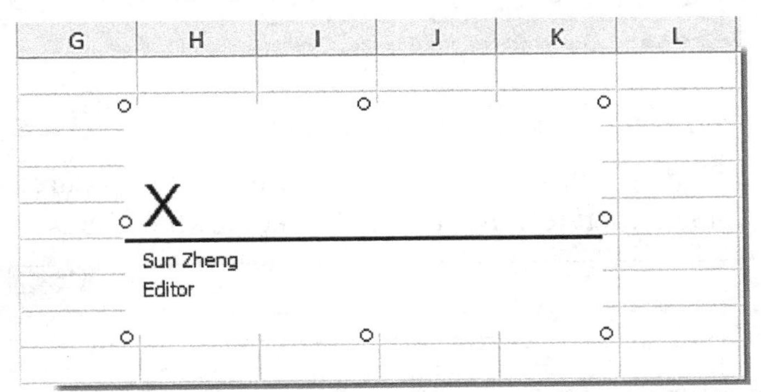

Double-click any graphical object in the signature line, to bring up any "Have the Digital ID dialogue box where one can choose the sort of ID one may need. We'll look at this and see if we can create the digital ID. Please note the following screenshot:

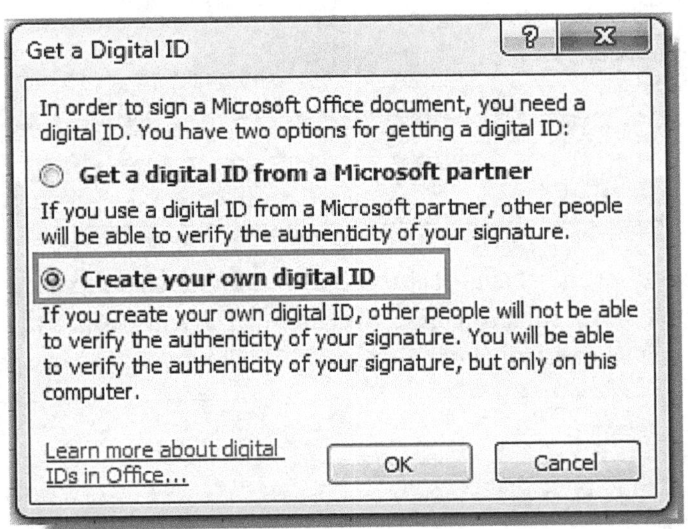

Please note that Excel 2k13-2k21 will not enable anyone to create the first digital ID. You can only do this by pressing the Yes icon in the Get some other Digital ID button and purchasing or installing the Digital ID. Otherwise, press the No option to exit. Also, take a look at the screengrab below.

Touch OK, then type the information into some other textbox to develop that Web ID dialogue, then click Create. Take a look at this screenshot:

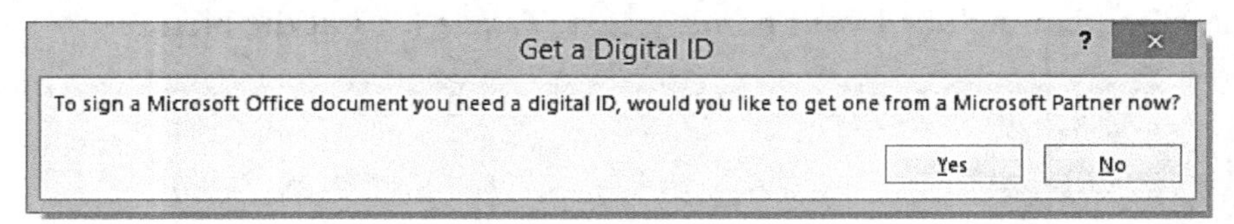

From the Signing tab, select and edit the first-ever signature image by pressing Choose Picture. Also, take a look at the screenshots.

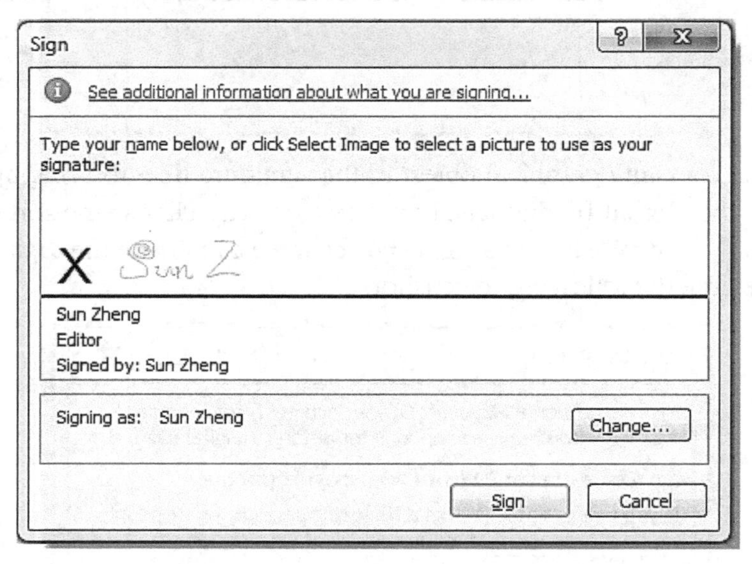

Enable the Get the Digital ID toolbar by double-clicking the signature text graphic item, then selecting the kind of ID you want. I'm checking to see if you've created a current digital ID. Please see the image below.

E	F	G	H	I	J	K
High	**Low**					
257	58					7/2/2014
390	302					
336	343		X			
71	266					
453	131		Sun Zheng			
465	107		Editor			
368	233		Signed by: Sun Zheng			
74	408					

Conclusion

The fresh Excel versions have all one needs to get started and then became a professional, with various fantastic features. MS Excel recognizes patterns and organizes data and thus saves your time. Create spreadsheets efficiently from templates or from scratch, then perform calculations with modern features. Through Microsoft 365 on Excel using mobile4, desktop, & web2 files, you could swap the workbooks with others while still working on the fresh editions having real-time synchronization to allow you to have the work done quickly.

It's both basic and advanced applications that could be used in almost any business environment. The Excel spreadsheet allows you to quickly & easily create, view, update, and share data with others. While reading and editing excel files attached to emails, you can create spreadsheets, data analyses, data charts, budgets, and much more. As you've become more familiar with different concepts, you should know the new resources & functionality that Excel provides to its users. The truth is that using Excel features, you can meet just about personal or business desires. What you need to do is invest your time and expand your knowledge.

A learning curve for improving certain skills will be long, but with practice and time, you will realize that things become second nature. After all, a man becomes perfect through practice.

CPSIA information can be obtained
at www.ICGtesting.com
Printed in the USA
BVHW061345160621
609724BV00002B/122